UNITED NATIONS ECONOMIC COMMISSION FOR EUROPE

PORCINE MEAT CARCASES AND CUTS

2006 EDITION

United Nations
New York and Geneva, 2008

NOTE

Working Party on Agricultural Quality Standards

The commercial quality standards of the UNECE Working Party on Agricultural Quality Standards help facilitate international trade, encourage high-quality production, improve profitability and protect consumer interests. UNECE standards are used by Governments, producers, traders, importers and exporters, and other international organizations, and cover a wide range of agricultural products, including fresh fruit and vegetables, dry and dried produce, seed potatoes, meat, cut flowers, eggs and egg products. For more information on UNECE agricultural standards, please visit our website <www.unece.org/trade/agr>.

This present edition of the Standard for Porcine Meat – Carcases and Cuts is based on document ECE/TRADE/C/WP.7/2006/14.

The designations employed and the presentation of the material in this publication do not imply the expression of any opinion whatsoever on the part of the United Nations Secretariat concerning the legal status of any country, territory, city or area or of its authorities, or concerning the delimitation of its frontiers or boundaries. Mention of company names or commercial products does not imply endorsement by the United Nations.

Please contact us at the following address with any comments or enquiries:

Agricultural Standards Unit
Trade and Timber Division
United Nations Economic Commission for Europe
Palais des Nations

CH-1211 Geneva 10, Switzerland
Tel: +41 22 917 1366
Fax: +41 22 917 0629
e-mail: agristandards@unece.org

To obtain copies of publications, please contact

United Nations Publications
Marketing, Sales and Licensing
Palais des Nations
CH-1211 Geneva 10, Switzerland
Tel: +41 22 917 32 23
Fax: +41 22 917 00 27
e-mail: unpubli@unog.ch

ECE/TRADE/369

UNITED NATIONS PUBLICATION
Sales No. E.07.II.E.1
ISBN 978-92-1-116953-9
ISSN 1810-1917

Preface

One of the principal goals of the United Nations Economic Commission for Europe (UNECE) is to promote greater economic integration of its members. As one activity for achieving this goal, UNECE provides a forum for Governments to develop internationally harmonized standards that:

- Facilitate fair international trade and prevent technical barriers to trade
- Define a common trading language for sellers and buyers
- Promote a high quality, sustainable production
- Create market transparency for buyers and consumers.

UNECE began work on standards for perishable produce in 1949. Today, close to 100 internationally harmonized, commercial quality standards have been developed for different agricultural produce: fresh fruit and vegetables, dry and dried produce, potatoes (early, ware and seed), eggs and egg products, meat and cut flowers.

Issues of commercial quality that have implications for international trade can be discussed in different specialized groups, and assistance is offered to countries that are interested in implementing UNECE standards (e.g. training workshops and seminars).

For each standard it is the aim to involve all interested parties in the work (members and non-members of UNECE, international governmental and non-governmental organizations) and to come to a consensus acceptable to all. It is a sign of the quality of UNECE standards that they have served as a basis for many European Union, Codex Alimentarius and OECD standards.

The UNECE standards for meat occupy a special place because of the complexity of the subject: a large number of product options can be specified by the buyer and the quality of the final product depends to a large extent on the way the meat is cut.

The standards offer, for the first time, internationally agreed specifications written in a consistent, detailed and accurate manner using anatomical names to identify cutting lines. Comprehensive colour photographs and diagrams are included to facilitate practical application of the standards.

The standards also define a product code allowing all relevant information to be combined in 20 digits. In developing this code, UNECE cooperated closely with GS1 International, a not-for-profit private-sector organization that supports supply chain systems with globally unique identification codes and electronic communications (e.g. bar codes).

The standardization of the trading language is the foundation which allows the meat industry to adopt modern data transfer methods and streamline the flow of information and products throughout the supply chain.

I hope that the new edition of the UNECE Standard for Porcine Meat – Carcases and Cuts will contribute substantially to the facilitation of fair international trade.

Marek Belka
Executive Secretary
United Nations Economic Commission for Europe

Acknowledgements

UNECE would like to thank all delegations who have contributed to the creation of the UNECE Standard for Porcine Meat - Carcases and Cuts: Australia, Bolivia, Brazil, China, GS1 International, European Union, France, Netherlands, New Zealand, Poland, Russian Federation, and United States of America. UNECE would also like to thank in particular the delegations of Australia and the United States for preparing the draft version of this standard and for providing the photographs.

Contents

UNECE STANDARD

PORCINE MEAT

CARCASES AND CUTS

1. INTRODUCTION

1.1 UNECE standards for meat products

The purpose of UNECE standards for meat products is to facilitate trade by recommending an international language for use between buyer and seller. The language describes meat items commonly traded internationally and defines a coding system for communication and electronic trade. As the texts will be updated regularly, meat industry members who believe that additional items are needed or that existing items are inaccurate or no longer being traded are encouraged to contact the UNECE secretariat.

The text of this publication has been developed under the auspices of the UNECE Specialized Section on Standardization of Meat. It is part of a series of standards which UNECE has developed or is planning to develop.

The following table contains the species for which UNECE standards exist or are being developed and their code for use in the UNECE meat code (see section 4).

For further information please visit the UNECE website at <www.unece.org/trade/agr>.

Annex I contains a description of the codification system, which includes a specific application identifier for the implementation of the UNECE code.

Species	Species code (data field 1)
Bovine (Beef)	10
Bovine (Veal)	11
Porcine (Pork)	30
Ovine (Sheep)	40
Caprine (Goat)	50
Llama	60
Alpaca	61
Chicken	70
Turkey	71

1.2 Scope

This Standard recommends an international language for raw (unprocessed) pork (*porcine)* carcases and cuts marketed as fit for human consumption. It provides purchasers with a variety of options for meat handling, packing and conformity assessment that conform to good commercial practice for meat and meat products intended to be sold in international trade.

To market porcine carcases and cuts across international borders, the appropriate legislative requirements of food standardization and veterinary control must be complied with. The Standard does not attempt to prescribe those aspects, which are covered elsewhere. Throughout the Standard, such provisions are left for national or international legislation, or requirements of the importing country.

The Standard contains references to other international agreements, standards and codes of practice that have the objective of maintaining the quality after dispatch and of providing guidance to Governments on certain aspects of food hygiene, labelling and other matters that fall outside the scope of this Standard. *Codex Alimentarius Commission Standards, Guidelines, and Codes of Practice* should be consulted as the international reference for health and sanitation requirements.

1.3 Application

Contractors are responsible for delivering products that comply with all contractual and specification requirements and are advised to set up a quality control system designed to assure compliance.

For assurance that items comply with these detailed requirements, buyers may choose to use the services of an independent, unbiased third-party to ensure product compliance with a purchaser's specified options. The standard includes illustrative photographs of carcases and selected commercial parts/cuts to make it easier to understand the provisions.

1.4 Adoption and publication history

The first edition of this standard was published in 1998. This second edition aligns the standard with other UNECE standards for meat and was adopted by the Working Party on Agricultural Quality Standards at its 62nd session in 2006.

In the second edition, agreed by the Specialized Section on Standardization of Meat at its 15th session in 2006, (see ECE/TRADE/C/WP.7/2006/14) a number of editorial changes were made to the original text adopted. The standard is now presented in five chapters in order to align it with the other standards. This alignment also included a reordering of the data fields in the porcine code and minor corrections to the carcases and cuts descriptions.

UNECE standards for meat undergo a complete review three years after publication. Following the review, new editions are published as necessary. Changes requiring immediate attention are published on the UNECE website at <www.unece.org/trade/agr/standards.htm>.

2. Minimum Requirements

All meat must originate from animals slaughtered in establishments regularly operated under the applicable regulations pertaining to food safety and inspection.

Carcases/cuts must be:

- Intact, taking into account the presentation
- Free from visible blood clots, or bone dust
- Free from any visible foreign matter (e.g. dirt, wood, plastic, metal particles[1])
- Free of offensive odours
- Free of obtrusive bloodstains
- Free of unspecified protruding or broken bones
- Free of contusions having a material impact on the product
- Free from freezer-burn[2].

Cutting, trimming and boning of cuts shall be accomplished with sufficient care to maintain cut integrity and identity, and avoid scores in the lean. Ragged edges shall be removed close to the lean surfaces. Except for cuts that are separated through natural seams, all cross-sectional surfaces shall form approximate right angles with the skin surface. Minimal amounts of lean, fat, or bone may be included on a cut from an adjacent cut. For boneless cuts, all bones, cartilage and visible surface lymph glands shall be removed.

3. Purchaser-Specified Requirements

The following subsections define the requirements that can be specified by the purchaser together with the codes to be used in the UNECE porcine code (see section 4).

3.1 Additional requirements

Additional purchaser-specified requirements, which are either not accounted for in the code (e.g. if code 9 "other" is used) or that provide additional clarification on the product or packing description shall be agreed between buyer and seller and be documented appropriately.

3.2 Species

The species code for porcine in data field 1 as defined in section 1.1 is 30.

3.3 Product/cut

The porcine cuts listed in this document are recommendations only. Different cuts of meat will be added or deleted as necessary as updates of this document evolve. Many of these cuts are traded internationally under the auspices of more than one trade name. The objective of using an harmonized codification system (see annex I) will facilitate the use of this document.

The four-digit product code in data field 2 is defined in section 5.

[1] When specified by the purchaser, meat items will be subject to metal particle detection.
[2] Freezer-burn is localized or widespread areas of irreversible surface dehydration indicated, in part or all, by changes from original colour (usually paler), and / or tactile properties (dry, spongy).

3.4 Refrigeration

Meat may be presented chilled, frozen or deep-frozen. Depending on the refrigeration method used, tolerances for product weight to be agreed between buyer and seller. Ambient temperatures throughout the supply chain should be such as to ensure uniform internal product temperatures as follows:

Refrigeration code (data field 4)	Category	Description
0	Not specified	
1	Chilled	Internal product temperature maintained at not less than -1.5° C or more than +7° C at any time following the post-slaughter chilling process
2	Frozen	Internal product temperature maintained at not exceeding -12° C at any time after freezing
3	Deep-frozen	Internal product temperature maintained at not exceeding -18° C at any time after freezing
4 - 8	Codes not used	
9	Other	

3.5 Production history

3.5.1 Traceability

The requirements concerning production history that may be specified by the purchaser require traceability systems to be in place. Traceability requires a verifiable method of identification of porcine animals, carcases, cartons and cuts at all stages of production. Traceability records must be able to substantiate the claims being made and the conformity of the procedures must be certified in accordance with the provisions concerning conformity-assessment requirements in section 3.12.

3.5.2 Porcine category

Porcine category code (data field 5)	Category	Description
0	Not specified	No category specified
1	Hog/Barrow	Castrated male porcine
2	Gilt	Female porcine, unfarrowed
3	Hog/Barrow and/or Gilt	Porcine
4	Sucker	Young porcine less than 15 kg (hot carcase weight), head-on
5	Boar	Mature intact porcine
6	Sow	Female porcine that has farrowed
7	Young pig	Young porcine less than 35 kg (hot carcase weight), head-on
8	Code not used	
9	Other	

3.5.3 Production system

The purchaser may specify a production system. In any case the production has to be in conformity with the regulations in force in the importing country. If no such regulation exists, the regulation of the exporting country shall be used.

Production system code (data field 6)	Category	Description
0	Not specified	No system specified
1	Indoors	Production methods that are based on indoor housing
2	Outdoors	Production methods that are based on outdoor housing for part of their lives
3	Organic	Production methods that conform to the legislation of the importing country concerning organic production
4 - 8	Codes not used	
9	Other	Any other production system agreed between buyer and seller

3.5.4 Feeding system

The purchaser may specify a feeding system. In any case the feeding has to be in conformity with the regulations in force in the importing country. If no such regulation exists, the feeding system shall be agreed between buyer and seller.

Feeding system code (data field 7)	Description
00	Not specified
01	Conventional
02 - 09	Codes not used
10	FM free
11	FM & IAO free
12	FM, IAO & GP free
13	FM, IAO, GP & GMO free
14	FM & GP free
15	FM, GP & GMO free
16	FM & GMO free
17 - 29	Codes not used
30	IAO free
31	IAO & GP free
32	IAO & GMO free
33	IAO, GP & GMO free
34 - 49	Codes not used
50	GP free
51	GP & GMO free
52 - 59	Codes not used
60	GMO free
61 - 98	Codes not used
99	Other

The definitions of the terms below have to be in conformity with the legislation of the importing country:

FM free	Free from fish meal
IAO free	Free from ingredients of animal origin
GP free	Free from growth promoters
GMO free	Free of products derived from genetically modified organisms.

3.5.5 Slaughter system

The purchaser may specify a slaughter system. In any case the slaughter has to be in conformity with the regulations in force in the importing country. If no such regulation exists, the slaughter system shall be agreed between buyer and seller.

Slaughter system code (data field 8)	Category	Description
0	Not specified	
1	Specified	Slaughter system specified as agreed between buyer and seller
2 - 8	Codes not used	
9	Other	Any other authorized method of slaughter must be agreed between buyer and seller

3.5.6 Post-slaughter system

The purchaser may specify a post-slaughter system. In any case the post-slaughter has to be in conformity with the regulations in force in the importing country. If no such regulation exists, the post-slaughter system shall be agreed between buyer and seller.

Post-slaughter processing codes (data field 9)	Category	Description
0	Not specified	
1	Specified	Post-slaughter system specified as agreed between buyer and seller
2 - 9	Codes not used	

NOTE 1: Spinal cord removal, individual market requirements will have specific regulations governing the removal of the spinal cord, nervous and lymphatic tissues, or other material. Regulations applicable to spinal cord removal will specify at what stage the carcase and/or cut must have the spinal cord removed. If required, there must be total removal.

NOTE 2: The following common post-slaughter processes, dressing specifications and chilling regimes, may be agreed between buyer and seller. These requirements are not included in the porcine-specific coding.

3.6 Fat limitations and evaluation of fat thickness in certain cuts

3.6.1 Fat thickness

The purchaser can specify the maximum fat thickness of partially skinned or skinless carcases, sides and cuts. Allowable fat limitations are as follows:

Fat thickness code (data field 10)	Category
0	Not specified
1	Peeled, denuded, surface membrane removed
2	From 0 to 5 mm fat thickness
3	From 6 mm to 12 mm fat thickness
4 - 8	Codes not used
9	Other

NOTE : Location of fat measurements on carcases to be agreed by buyer and seller (e.g. rib sites). For information on the calculation of the percentage of lean, see section 3.8.

3.6.2 Trimming

Trimming of external fat shall be accomplished by smooth removal along the contour of underlying muscle surfaces. Bevelled fat edges alone do not substitute for complete trimming of external surfaces when required. Fat thickness requirements may apply to surface fat (subcutaneous and/or exterior fat in relation to the item) and seam (intermuscular) fat as specified by the purchaser. Two definitions are used to describe fat trim limitations:

- Maximum fat thickness at any one point. Evaluated by visually determining the area of a cut which has the greatest fat depth, and measuring the thickness of the fat at that point.
- Average (mean) fat thickness. Evaluated by visually determining and taking multiple measurements of the fat depth of areas where surface fat is evident only. Average fat depth is determined by computing the mean depth in those areas.

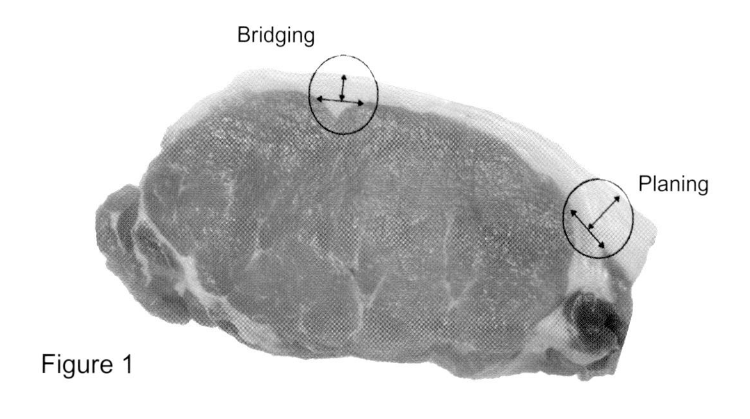

Figure 1

Actual measurements of fat thickness (depth) are made on the edges of cuts by probing or scoring the overlying surface fat in a manner that reveals the actual thickness and accounts for any natural depression or seam that could affect the measurement. When a natural depression occurs in a muscle, only the fat above the portion of the depression which is more than 19 mm (0.75 inch) in width is considered (known as bridging; see figure 1). When a seam of fat occurs between adjacent muscles, only the fat above the level of the involved muscles is measured (known as planing; see figure 1).

However, when fat limitations for peeled/denuded, surface membrane removed[3] are specified, the bridging method shall be used for evaluating fat above a natural depression in a muscle and fat occurring between adjacent muscles.

3.7 Porcine quality system

Porcine quality system code (data field 11)	Category	Description
0	Not specified	
1	Official standards	Quality classifications based on an official quality system of the exporting country
2	Company standards	Quality classifications based on sellers' quality system
3	Industry standards	Quality classifications based on an industry-wide quality system
4 - 8	Codes not used	
9	Other	Other quality classifications agreed between buyer and seller

NOTE : Any system should meet or exceed the official quality requirements of the consuming country.

3.8 Meat and fat colour, marbling and pH

Normally, lean meat and fat, depending on the specific species, demonstrates a characteristic colour and pH. Specific requirements regarding colour, marbling and pH if required need to be agreed between buyer and seller and are not provided for in the coding system.

The specified system requirements will be agreed upon between the buyer and seller. These quality systems may include, but are not limited to, percentage of lean product, marbling, lean colour and pH. These different quality standards are based on specifications developed by different countries, companies and/or industries.

[3] Peeled/denuded, surface membrane removed – When the surface membrane ("silver" or "blue tissue") is required to be removed (skinned), the resulting cut surface shall expose at least 90 per cent lean with remaining "flake" fat not to exceed 3 mm (0.125 inch) in depth.

3.9 Weight ranges of carcases and cuts

Weight range code (data field 12)	Category	Description
0	Not specified	
1	Specified	Range required
2 - 9	Codes not used	

NOTE : These weight ranges are not available for portion control, but rather a range to delineate the size of cuts being sold.

3.10 Packing, storage, and transport

3.10.1 Description and provisions

The primary packaging is the primary covering of a product and must be of food grade materials. The secondary packaging contains products packaged in their primary packaging. During storage and transport, the meat must be packaged to the following minimum requirements:

Carcases, split carcase sides and quarters
- Chilled, frozen or deep-frozen with or without packaging

Cuts - chilled
- Individually wrapped (I.W.)
- Bulk packaged (plastic or wax-lined container)
- Vacuum-packed (VAC)
- Modified atmosphere packaging (MAP)
- Other

Cuts - frozen / deep frozen
- Individually wrapped (I.W.)
- Bulk packaged (plastic or wax-lined container)
- Vacuum-packed (VAC)
- Other

The conditions of storage before dispatch and the equipment used for transportation shall be appropriate to the physical and, in particular, the thermal condition of the meat (chilled, chilled in a modified atmosphere, frozen, or deep-frozen) and shall be in accordance with the requirements of the importing country. Attention is drawn to the provisions of the *UNECE Agreement on the International Carriage of Perishable Foodstuffs and on the Special Equipment to be Used for Such Carriage (ATP)* (ECE/TRANS/165).

3.10.2 Packing code

Packing code (data field 13)	Category
0	Not specified
1	Carcases, split carcase sides and quarters - without packaging
2	Carcases, split carcase sides and quarters - with packaging
3	Cuts - individually wrapped (I.W.)
4	Cuts - bulk packaged (plastic or wax-lined container)
5	Cuts - vacuum-packed (VAC)
6	Cuts - modified atmosphere packaging (MAP)
7	Layer packed with plastic or wax-lined dividers
8	Code not used
9	Other

3.11 Labelling information to be mentioned on or affixed to the marketing units of meat

All labelling information must be verifiable (see also section 3.5.1).

3.11.1 Mandatory information

Without prejudice to national requirements of the importing countries, the following table contains information that must be listed on product labels.

- For carcase sides and quarters, the mandatory information must be affixed to the product (stamped and/or tagged).
- For packaged cuts, the mandatory information must be listed on the shipping container.

Labelling information	Unpackaged carcases, quarters and cuts	Packaged or package meat
Health stamp	✗	✗
Slaughter number or batch number	✗	✗
Packaging date		✗
Name of the product		✗
Use-by date, as required by each country		✗
Storage conditions (see section 3.4 Refrigeration)		✗
Appropriate identification of packer, processor or retailer		✗ [4]
Quantity (number of pieces)		✗ [4]
Net weight		✗ [4]

[4] This information can also be provided in accompanying documentation.

3.11.2 Additional information

Additional information may be listed on product labels as required by the importing country's legislation, or at the buyer's request, or as chosen by the processor. If listed, such product claims must be verifiable (see also section 3.5.1).

Examples of such product claims include the following:

- Country of birth
- Country(ies) of raising
- Country of slaughter
- Country(ies) of processing/cutting
- Country(ies) of packing
- Country of origin. In this Standard the term "country of origin" is reserved to indicate that birth, raising, slaughter, processing/cutting and packing have taken place in the same country.
- Characteristics of the livestock, production and feeding systems
- Slaughter date
- Slaughter and post-slaughter systems
- Processing/packaging date
- Quality/grade/classification
- pH, lean and fat colour.

3.12 Provisions concerning conformity-assessment requirements

The purchaser may request third-party conformity assessment of the product's quality/grade/ classification, purchaser-specified options of the Standard, and/or animal identification. Individual conformity assessments or combinations may be selected as follows:

Quality/grade/classification conformity assessment (quality): a third party examines and certifies that the product meets the quality level requested. The name of the third-party certifying authority and quality grade standard to be used must be designated as noted in 3.1.

Trade standard conformity assessment (trade standard): a third party examines and certifies that the product meets the purchaser-specified options as specified in this trade standard, except for quality level. The name of the third-party certifying authority must be designated as noted in 3.1. Optionally, the purchaser may indicate specific purchaser-specified options to be certified after the name of the third-party certifying authority.

Porcine or batch identification conformity assessment (porcine/batch ID): a third party certifies that the product meets specified requirements. The name of the third-party certifying authority and the requirements must be designated as noted in 3.1.

Conformity assessment code (data field 14)	Category
0	Not specified
1	Quality/grade/classification (quality) conformity assessment
2	Trade standard conformity assessment
3	Porcine/batch identification (porcine/batch ID) conformity assessment
4	Quality and trade standard conformity assessment
5	Quality and porcine/batch ID conformity assessment
6	Trade standard and porcine/batch ID conformity assessment
7	Quality, trade standard, and porcine/batch ID conformity assessment
8	Code not used
9	Other

4. UNECE Code for Purchaser Requirements for Porcine Meat

4.1 Definition of the code

The UNECE code for purchaser requirements for porcine meat has 14 fields and 20 digits (2 digits not used) and is a combination of the codes defined in sections 3 and 5.

Field no.	Name	Section	Code Range
1	Species	3.2	00 - 99
2	Product/cut	5	0000 - 9999
3	Field not used	-	00 - 99
4	Refrigeration	3.4	0 - 9
5	Category	3.5.2	0 - 9
6	Production system	3.5.3	0 - 9
7	Feeding system	3.5.4	00 - 99
8	Slaughter system	3.5.5	0 - 9
9	Post-slaughter system	3.5.6	0 - 9
10	Fat thickness	3.6.1	0 - 9
11	Quality	3.7	0 - 9
12	Weight range	3.9	0 - 9
13	Packing	3.10.2	0 - 9
14	Conformity assessment	3.12	0 - 9

4.2 Example

The following example describes a chilled, vacuum-packed, porcine leg long cut (style 1) of a specified weight range, trimmed to less than 5 mm fat thickness, from a hog/barrow raised in an indoor production system with a conventional feeding system, slaughtered and processed under specified requirements, with a company standard quality system applied.

This item has the following UNECE porcine code: **30401300111011122150**

No.	Name	Requirement	Code Value
1	Species	Porcine	30
2	Product/cut	Leg long cut	4013
3	Field not used	-	00
4	Refrigeration	Chilled	1
5	Category	Hog/barrow	1
6	Production system	Indoors	1
7	Feeding system	Conventional	01
8	Slaughter system	Specified	1
9	Post-slaughter system	Specified	1
10	Fat thickness	From 0 to 5 mm fat thickness	2
11	Quality	Company standards	2
12	Weight range	Specified	1
13	Packing	Cuts-vacuum-packed	5
14	Conformity assessment	Not specified	0

5. Carcases and Cuts Descriptions

5.1 Multilingual index of products

Item	English	Page	French	Russian
4000	Full carcase	22	Carcasse entière	Цельная туша
4001	Carcase side	22	Demi-carcasse	Полутуша
4002	Carcase side – Block ready (3–way)	23	Demi-carcasse prête à la découpe (trois coupes)	Полутуша для приготовления мясных блоков (3-составная)
4003	Carcase side – Block ready (3–way-special trim)	23	Demi-carcasse prête à la découpe (trois coupes – découpe spéciale)	Полутуша для приготовления мясных блоков (3-составная специальной разделки)
4004	Carcase side – Block ready (4–way-special trim)	24	Demi-carcasse prête à la découpe (quatre coupes – découpe spéciale)	Полутуша для приготовления мясных блоков (4-составная специальной разделки)
4009 – 4010	Hindquarter	25	Quartier arrière	Задняя четвертина
4011	Roasting pig, full	25	Porc à rôtir – entier	Целиковый поросенок для жарения
4012	Roasting pig, split	25	Porc à rôtir – demi	Разрубленный по позвоночнику поросенок для жарения
4013 – 4015	Leg long cut	26	Cuisse – coupe longue	Тазобедренный отруб длинный
4016 – 4018	Leg short cut	27	Cuisse – coupe courte	Тазобедренный отруб короткий
4021 – 4026	Forequarter	24	Quartier avant	Передняя четвертина
4029 – 4032	Shoulder - square cut	39	Quartier avant sans gorge et sans cotis	Лопаточная часть - квадратный отруб
4044	Shoulder outside	40	Épaule	Наружная часть лопатки
4045	Shoulder outside (3–way)	40	Épaule (trois pièces)	Наружная часть лопатки (3-составная)
4046 – 4049 4063	Shoulder inside	41	Échine	Внутренняя часть лопатки
4050 – 4055	Shoulder lower half (Shoulder-picnic)	41	Épaule – hachage	Нижняя часть лопатки (пикниковая лопатка)
4059 – 4062	Shoulder upper half (Butt or collar butt)	41	Échine palette	Верхняя часть лопатки (край или шейный край)
4069 – 4072	Middle	31	Milieu	Средняя часть
4079 – 4082	Belly	32	Poitrine	Грудо-реберный отруб с пашиной
4098 – 4101	Loin – centre cut	33	Longe – sans échine et sans pointe	Спинно-поясничный - центральный отруб
4102 – 4105	Semiboneless loin – centre cut	34	Longe – sans échine et sans pointe semi-désossée	Полуобваленный спинно-поясничный - центральный отруб
4108 – 4111	Loin – long (Blade removed)	33	Longe – sans palette et sans couenne	Спинно-поясничный отруб длинный (без лопаточной кости)
4113	Loin – long (4–way)	33	Longe (quatre pièces)	Спинно-поясничный отруб длинный (4-составной)

Item	English	Page	French	Russian
4130	Sirloin (rump)	34	Pointe	Верхняя часть тазобедренного отруба
4140 – 4147	Loin – long	32	Longe – avec couenne	Спинно-поясничный отруб длинный
4159	Loin riblets	34	Apophyses de filet (loin riblets)	Реберная часть спинно-поясничного отруба
4160	Belly ribs (Spare ribs)	35	Plat de poitrine (spare ribs)	Грудо-реберный отруб (ребра без поверхностного мяса)
4161	Back ribs (Loin ribs)	35	Plat de longe (loin ribs)	Верхняя часть реберного отруба (реберная часть корейки)
4162	Full rib plate	35	Plat de milieu	Реберный отруб (полная реберная пластина)
4163	St. Louis style ribs	35	Côtes – style St. Louis	Ребра разделки "сен-луи"
4164	Short ribs	36	Travers	Реберный край
4165 – 4167	Shoulder ribs	44	Cotis	Подлопаточные ребра
4170	Hock shoulder	44	Jarret avant	Рулька
4172	Hock leg (Ossobucco)	31	Jarret arrière (osso-bucco de porc)	Рулька ("оссобукко")
4175	Fore feet (Trotter)	45	Pieds avant	Передняя ножка
4176	Hind feet (Trotter)	31	Pieds arrière	Задняя ножка
4180	Shoulder (M. pectoralis)	43	Bateau (M. pectoralis)	Лопатка (M. pectoralis)
4181	Shoulder (M. teres major)	43	Dessus de palette (M. teres major)	Лопатка (M. teres major)
4182	Shoulder (M. serratus ventralis)	43	Persillé (M. serratus ventralis)	Лопатка (M. serratus ventralis)
4183	Shoulder (Cushion)	43	Macreuse	Лопатка (подушка)
4200	Leg long cut (boneless)	28	Cuisse – coupe longue (désossée)	Тазобедренный (бескостный) отруб длинный
4240	Collar butt – special trim (Butt or collar butt – special trim)	42	Échine palette – parage spécial	Шейный отруб (особая разделка) (край или шейный край)
4241	Shoulder inside (boneless)	42	Échine (désossée)	Внутренняя часть лопатки (бескостная)
4245	Shoulder upper half (boneless) (Butt or collar butt)	42	Échine palette (désossée)	Верхняя часть лопатки (бескостная) (край или шейный край)
4280	Tenderloin	38	Filet mignon	Вырезка
4290	Inside	29	Noix	Внутренняя часть тазобедренного отруба
4300	Outside (Outside trimmed or silverside)	28	Sous-noix (sous-noix parée ou noix à escalopes)	Наружная часть тазобедренного отруба (наружная часть тазобедренного отруба зачищенная или "сильверсайд")
4301	Outside eye	28	Semitendinosus (rond de gîte)	Наружная часть тазобедренного отруба (длинная мышца)
4305	Sirloin (rump) boneless	37	Pointe désossée	Верхняя часть тазобедренного отруба обваленная (кострец)

Item	English	Page	French	Russian
4310	Knuckle (tip)	29	Noix pâtissière	Боковая часть тазобедренного отруба (верхушка)
4311	3 – Way leg	29	Jambon (trois pièces)	3-составной тазобедренный отруб
4312	4 – Way leg	30	Jambon (quatre pièces)	4-составной тазобедренный отруб
4313	5 – Way leg	30	Jambon (cinq pièces)	5-составной тазобедренный отруб
4314	6 – Way leg	30	Jambon (six pièces)	6-составной тазобедренный отруб
4319 – 4322	Middle	36	Milieu désossé	Средняя часть
4329 – 4332	Belly (boneless)	38	Poitrine (désossée)	Грудо-реберный отруб с пашиной бескостный
4333	Belly (Flank on)	39	Poitrine (avec mouille)	Пашина
4335	Shoulder-picnic and belly	39	Poitrine hachage	Пикниковая лопатка и грудо-реберный отруб
4340 – 4343	Loin	36	Longe désossée	Спинно-поясничный отруб бескостный (эскалопная часть)
4350	Jowl	44	Gorge	Щековина
4360	Eye of shortloin	37	Filet	Поясничный отруб бескостный (филейная покромка)
4361	Eye of loin	37	Noix de longe	Спинной отруб бескостный (филейная вырезка)
4470	Trimmings	45	Parures	Обрезь
7680	Shoulder fat	45	Gras d'épaule	Лопаточный шпик
7685	Back fat	45	Bardière	Хребтовый шпик

5.2 Porcine side skeletal diagram

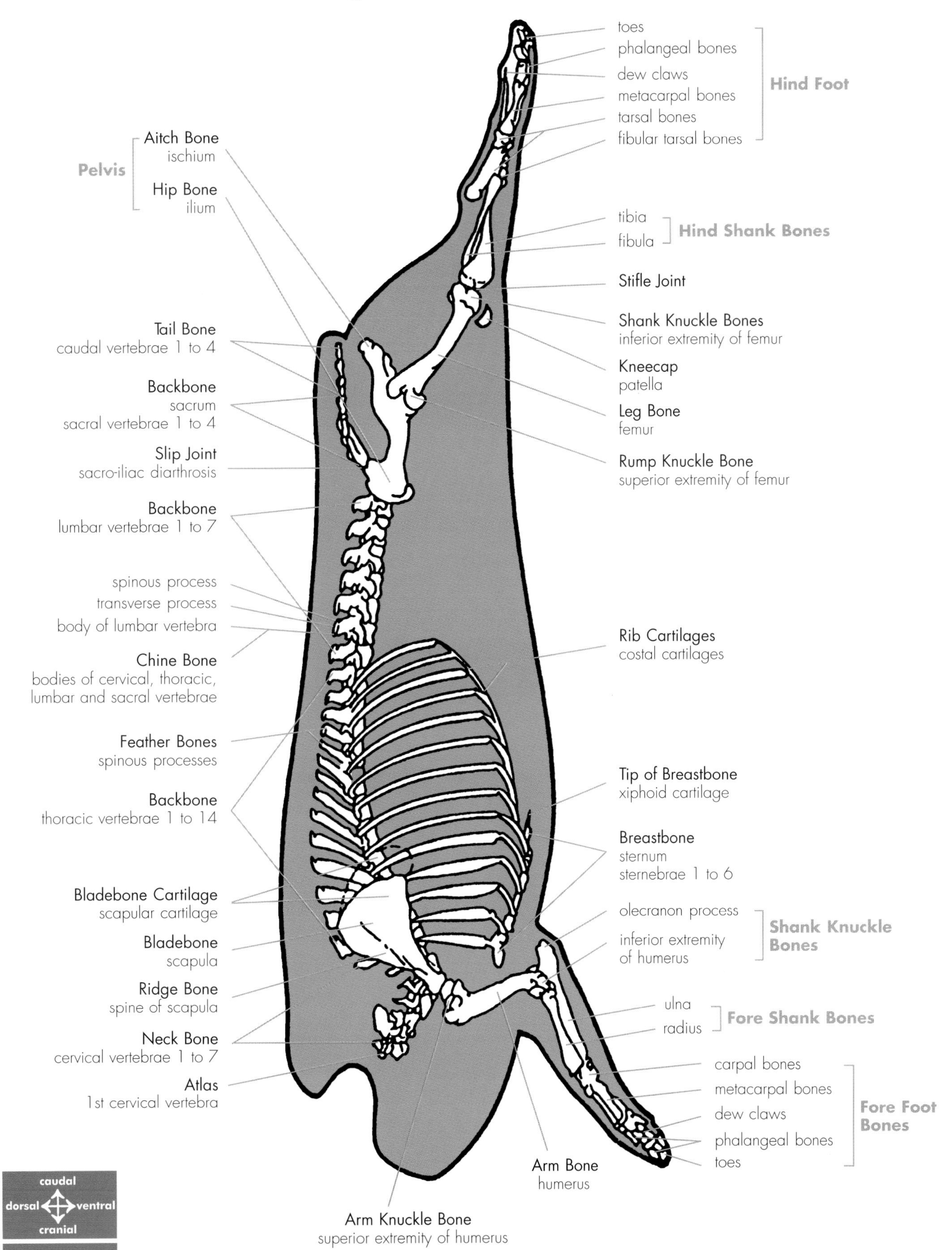

5.3 Standard porcine primal cuts flow chart

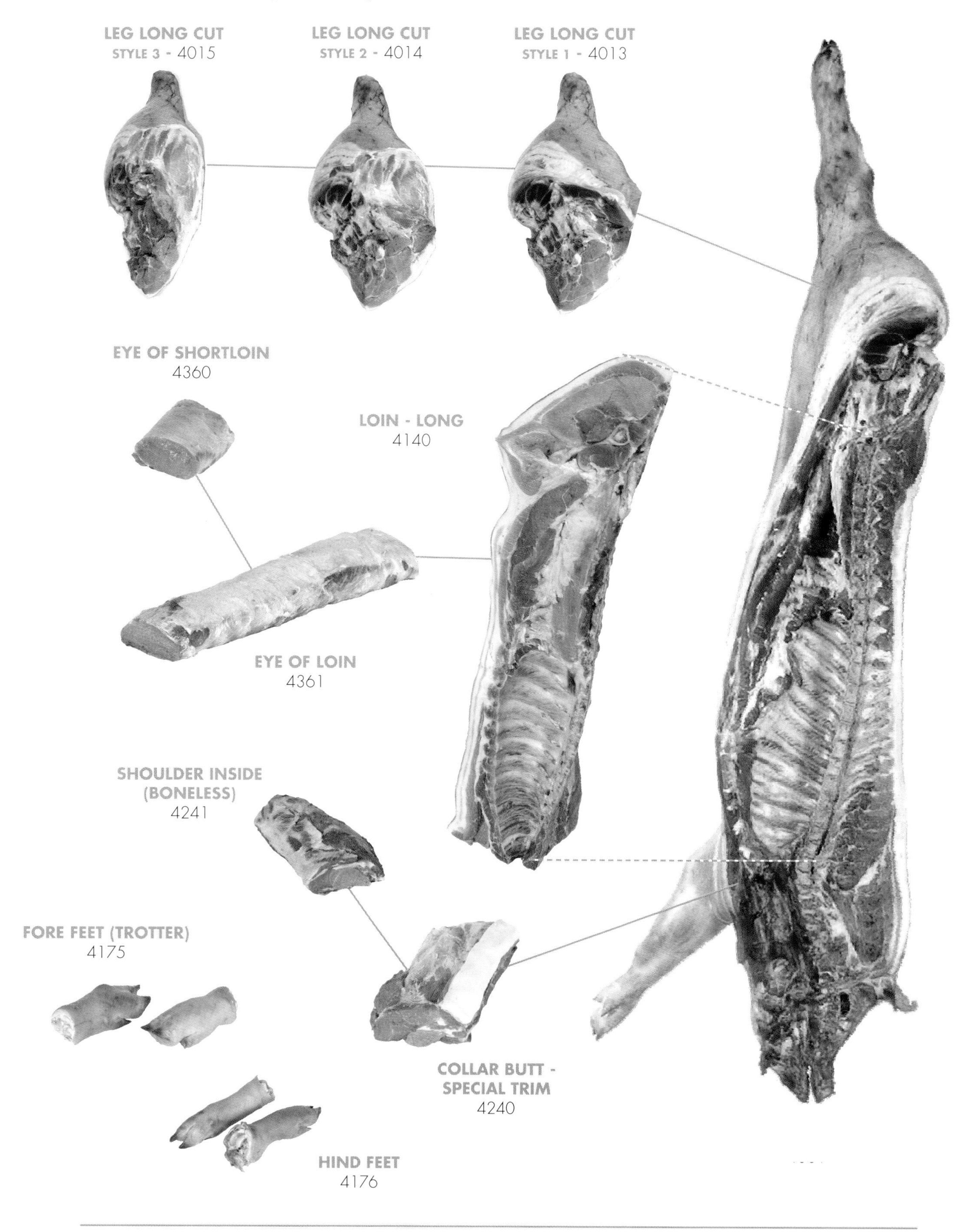

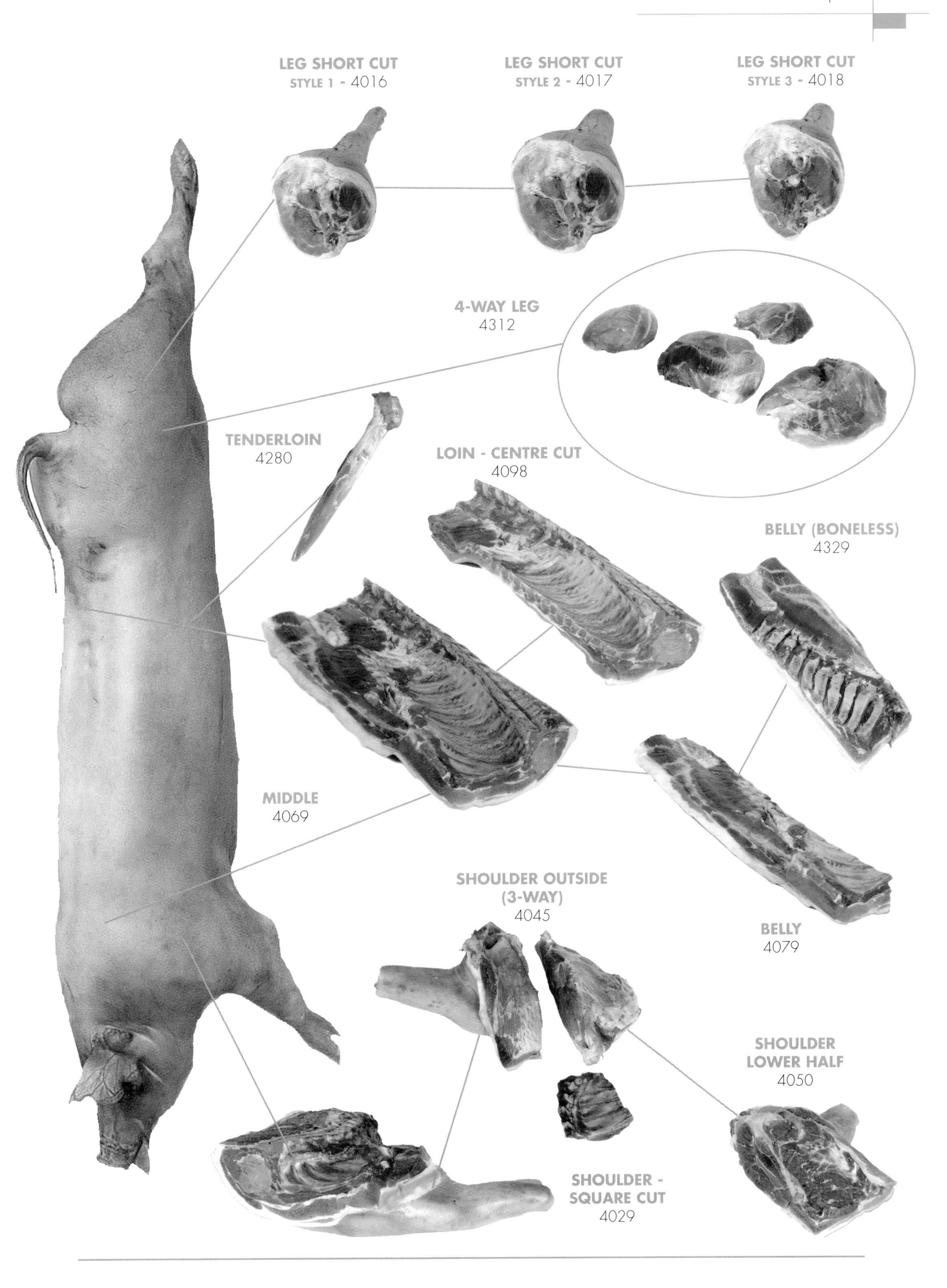
LEG SHORT CUT
STYLE 1 - 4016
LEG SHORT CUT
STYLE 2 - 4017
LEG SHORT CUT
STYLE 3 - 4018
4-WAY LEG
4312
TENDERLOIN
4280
LOIN - CENTRE CUT
4098
BELLY (BONELESS)
4329
MIDDLE
4069
SHOULDER OUTSIDE
(3-WAY)
4045
BELLY
4079
SHOULDER
LOWER HALF
4050
SHOULDER -
SQUARE CUT
4029

5.4 Porcine meat cuts

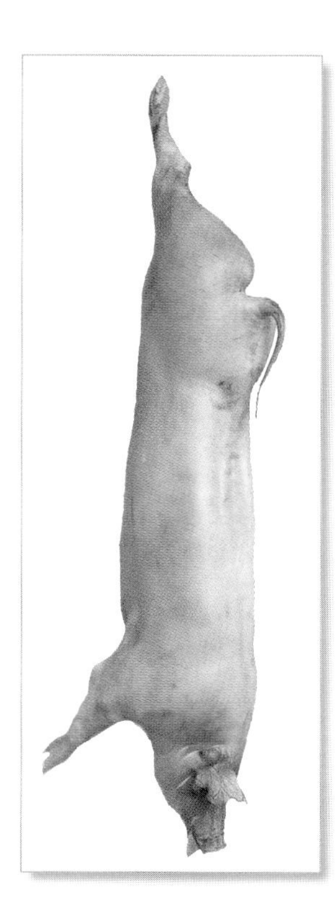

4000 FULL CARCASE

Full carcase includes all parts of the body skeletal musculature and bone, shall be dressed without the kidneys or other internal organs and shall be practically free of internal fat. The kidney, pelvic, heart and leaf fat may remain. There shall not be any objectionable scores on the outside of the carcase and, unless otherwise specified, the carcase shall be skin-on. Mutilated feet must be removed at the hock or upper knee joint (as applicable). Carcases with bloody "stuck" shoulders (caused by improper severing of the carotid artery) are not acceptable. The membranous portion of the diaphragm must be removed close to the lean, although the lean portion (and the membrane surrounding the lean portion) may remain if firmly attached to the carcase. Head, jowls and feet are retained unless otherwise specified. The tail is retained unless otherwise specified.

To be specified:

- Head removed
- Head and jowls removed
- Head removed and jowls retained
- Fore foot (trotter) removed
- Hind foot (trotter) removed
- Tail removed
- Diaphragm removed
- Pillar of diaphragm removed
- Flank fat adjacent to the leg removed
- Kidney, pelvic, heart, leaf fat removed

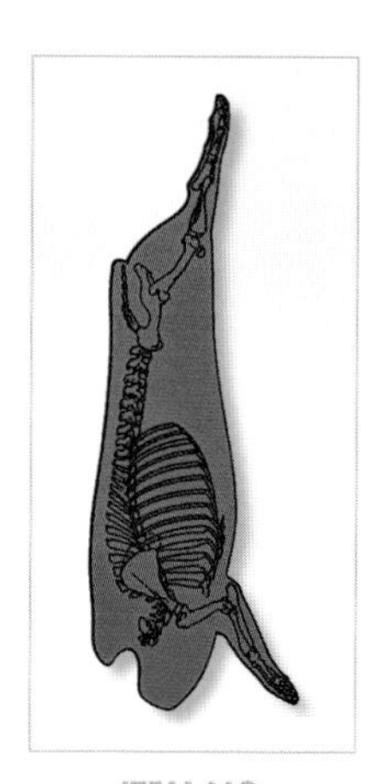

ITEM NO.
4000

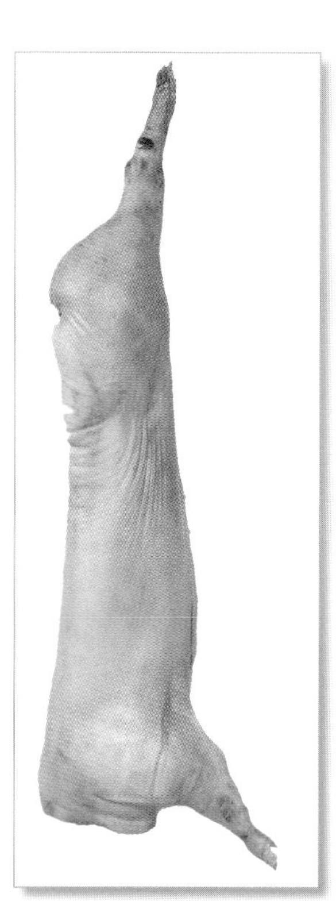

4001 CARCASE SIDE

Carcase side is prepared from the full carcase (item 4000). The carcase shall be split into reasonably uniform carcase sides by cutting lengthwise through the backbone so that the major muscles of the loin and shoulder are not scored and such that the spinal cord groove is evident throughout the length of the back bone. Jowl and hind foot are retained unless otherwise specified. The tail is removed. Head and fore foot are removed unless specified.

To be specified:

- Head retained
- Jowl removed
- Fore foot retained
- Hind foot removed

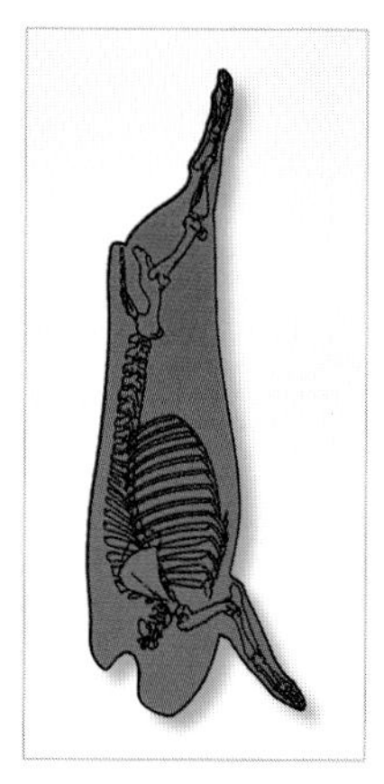

ITEM NO.
4001

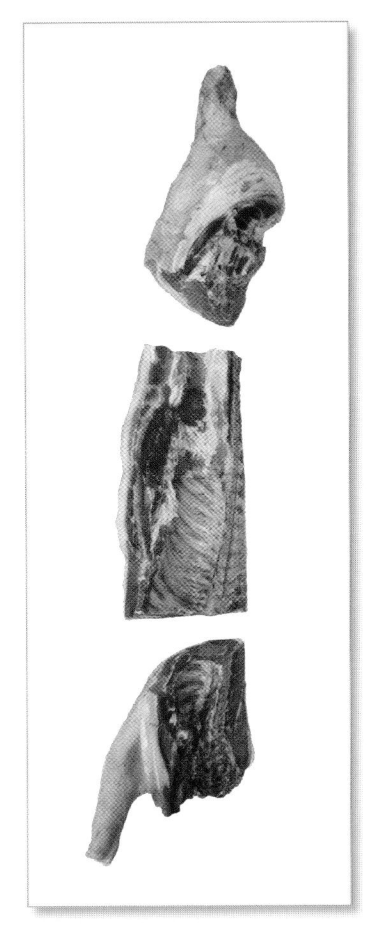

4002 CARCASE SIDE – BLOCK READY (3-WAY)

Carcase side – block ready consists of the same specifications as carcase side (item 4001). The carcase is cut in three sections approximately perpendicular to the length of the carcase. The cuts consist of a leg long cut (item 4013) removed by a cut through the vertebral column between the 6th and 7th lumbar vertebrae. The hind trotter (item 4176) is removed between the tarsus and metatarsus. The middle (item 4069) is removed from the forequarter along the specified rib. The forequarter (item 4024) is removed along the specified rib. The fore feet (trotter) (item 4175) is removed at the carpal joint. Jowl (item 4350) is removed.

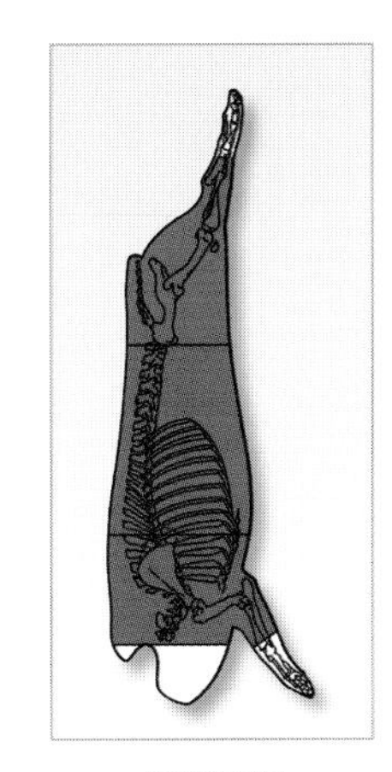

ITEM NO.
4002

4003 CARCASE SIDE – BLOCK READY (3-WAY-SPECIAL TRIM)

Carcase side – block ready (3-way-special trim) consists of the same carcase specifications as carcase side (item 4001). The carcase is cut in three sections. The cuts consist of a leg short cut (item 4016) with the hind trotter retained. The remaining trunk portion of the side is cut as a loin – long (item 4140) and shoulder picnic and belly (item 4335).

The loin long/shoulder picnic and belly separation point is made by a cut commencing at the cranial end starting at a specified distance from the vertebrae through the joint of the blade bone and humerus and parallel to the chine edge the full length of the loin to the tip of and including the extended muscle of the flank.

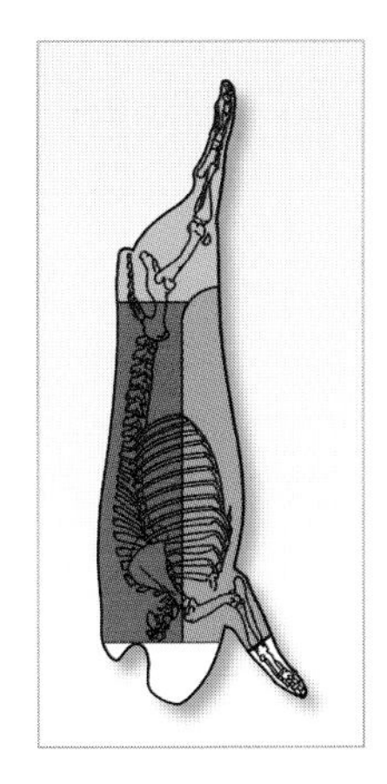

ITEM NO.
4003

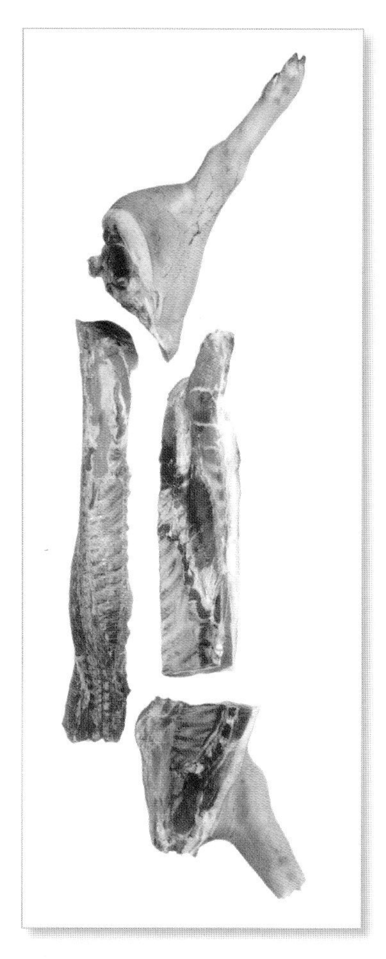

4004 CARCASE SIDE – BLOCK READY (4-WAY-SPECIAL TRIM)

Carcase side – block ready (4-way-special trim) consists of the same carcase specifications as carcase side (item 4001). The carcase is cut into four sections. The cuts consist of a leg short cut (item 4016) with the hind trotter retained. The remaining trunk portion of the side is cut as a loin - long (item 4140). The shoulder blade portion located over the loin is removed along the natural seam and attached to the forequarter portion. The ventral portion, shoulder lower half (item 4050) and belly – extended (item 4333) are separated by a straight cut along the specified rib.

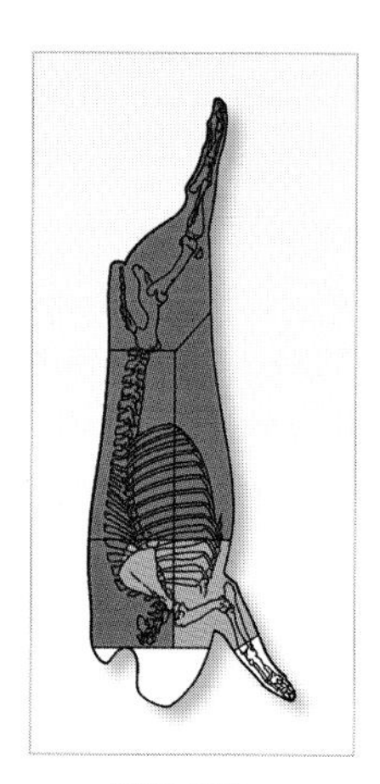

ITEM NO.
4004

4021 - 4026 FOREQUARTER

Forequarter is derived from a split carcase side (item 4001) by a straight cut through the vertebrae at a specified rib number, following the contour of the rib to the ventral portion of the belly. Foot, jowl and cervical/thoracic vertebrae and ribs/intercostals are retained unless otherwise specified.

To be specified:

- Removal of fore foot
- Removal of jowl
- Cervical/thoracic vertebrae removed
- Ribs/intercostal muscles removed

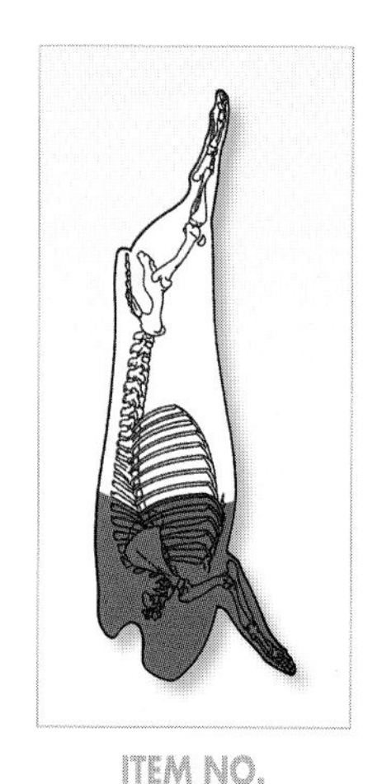

ITEM NO.
4021 (1-rib)
4022 (2-ribs)
4023 (3-ribs)
4024 (4-ribs)
4025 (5-ribs)
4026 (6-ribs)

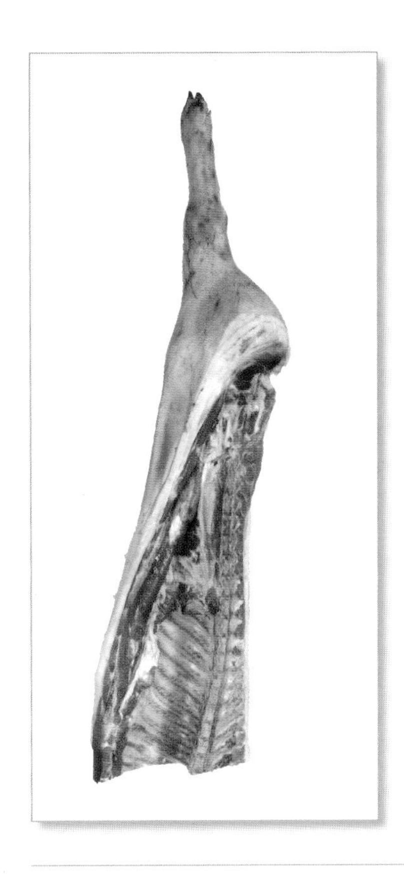

4009 - 4010 HINDQUARTER

Hindquarter is derived from a split carcase side (item 4001) by a straight cut through the vertebrae at a specified rib number, following the contour of the rib to the ventral portion of the belly. Alternative specifications shall be as agreed between buyer and seller. The diaphragm and foot are removed.

To be specified:

- Diaphragm removed
- Foot (trotter) removed

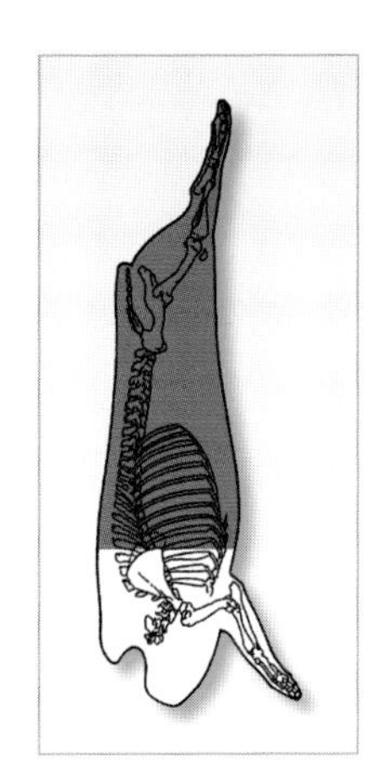

ITEM NO.
4009 (9-ribs)
4010 (8-ribs)

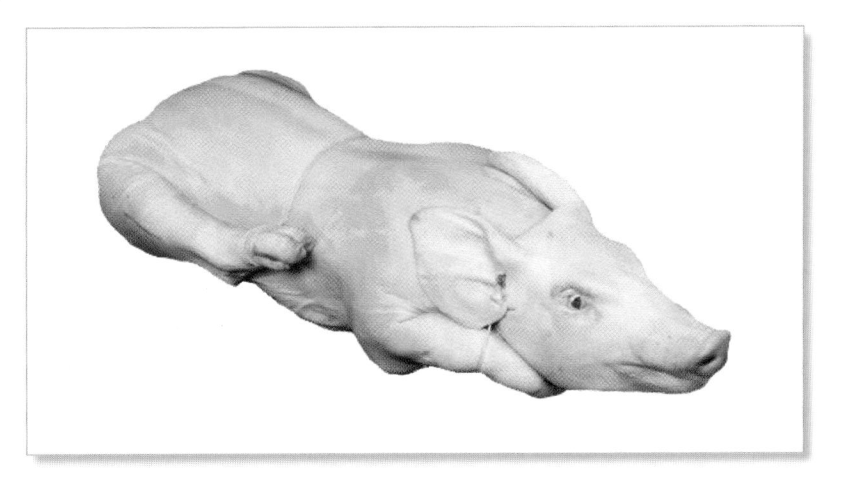

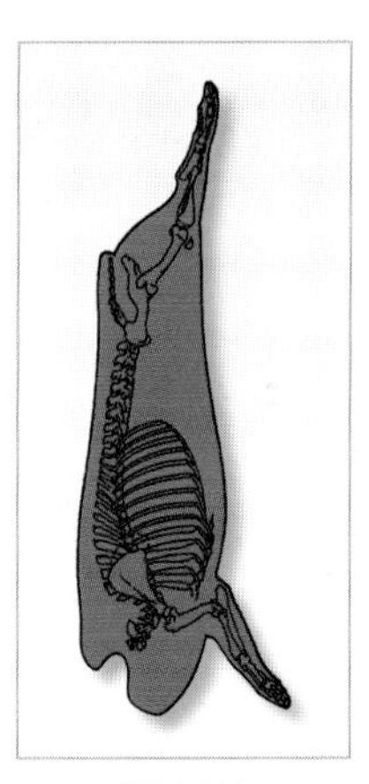

ITEM NO.
4011
4012

4011 ROASTING PIG, FULL

Roasting pig, full, has the head on, and may include the internal fat and the membranous portion of the diaphragm. If applicable other requirements may apply.

To be specified:

- Approximate weight

4012 ROASTING PIG, SPLIT

Roasting pig, split, consists of a roasting pig with the head remaining, and may include the internal fat and the membranous portion of the diaphragm. If applicable other requirements may apply.

To be specified:

- Approximate weight

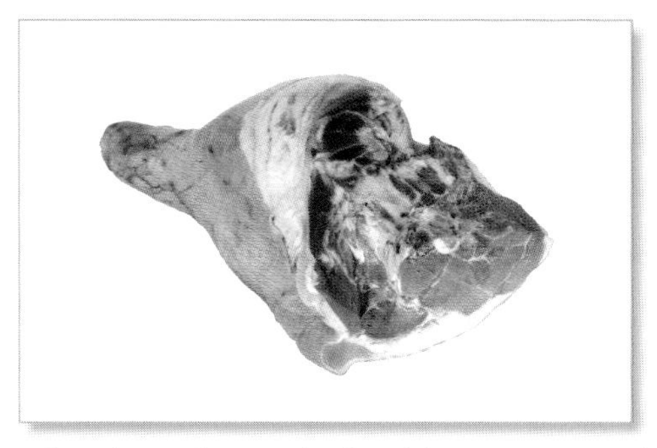

4013 LEG LONG CUT

Style 1

Leg long cut: style 1 is separated from the split carcase side (item 4001) by a straight cut approximately perpendicular to a line parallel to the vertebral column between the 6th and 7th lumbar vertebrae and passing through a point immediately anterior to the hip bone (ilium) and related cartilage. The hind foot is removed at the tarsal joint.

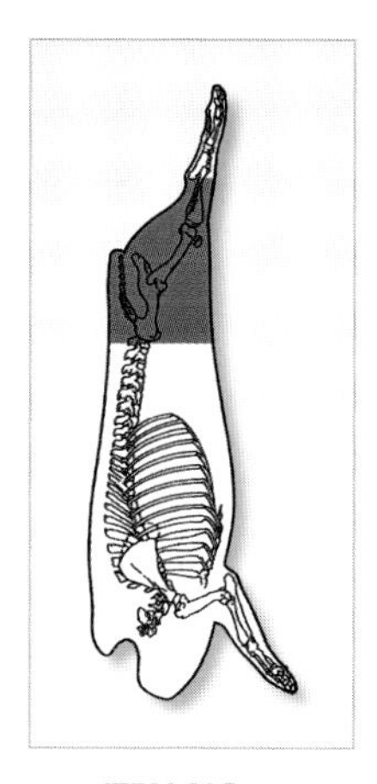

ITEM NO.
4013
4014
4015

To be specified:

- Foot retained

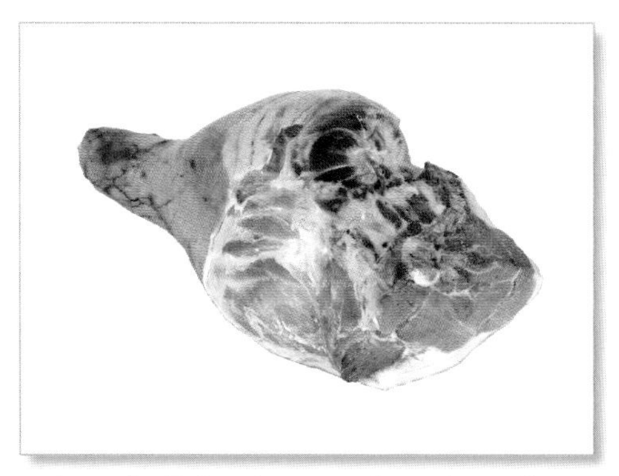

4014 LEG LONG CUT

Style 2

Leg long cut: style 2 is the same as style 1 except the tail (caudal) vertebrae, flank muscle (M. rectus abdominis), M. cutaneus trunci, and exposed lymph glands shall be removed. The skin and collar fat over the M. semimembranosus shall be smooth and well rounded such that the innermost curvature of the skin is trimmed back at least half the distance from the stifle joint to the posterior edge of the aitch bone. The skin overlying the medial side (inside) of the M. quadriceps femoris shall be removed and fat overlying the M. quadriceps femoris and pelvic area shall be removed close to the lean. The hind foot is removed at the tarsal joint.

To be specified:

- Foot retained

4015 LEG LONG CUT

Style 3

Leg long cut: style 3 is the same as style 2 except that flank muscles (M. rectus abdominis, M. obliquus internus abdominis, M. obliquus externus abdominis), vertebrae, hip bone along with overlying lean and fat, lean and fat overlying the quadriceps (fore cushion), M. psoas major and M. iliacus shall be removed. The ball of the femur shall be exposed. The hind foot is removed at the tarsal joint. The butt tenderloin shall be removed and skin is retained.

To be specified:

- Skin removed
- Foot (trotter) retained
- Butt tenderloin retained

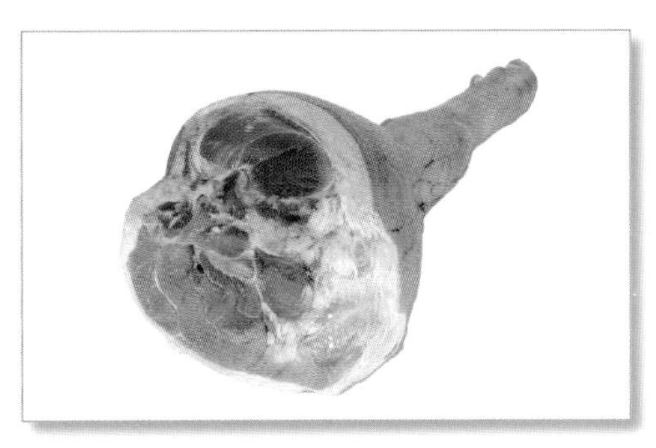

4016 LEG SHORT CUT

Style 1

Leg short cut: style 1 is separated from the split carcase side (item 4001) by a straight cut anterior to the quadriceps approximately perpendicular to a line parallel to the shank bones and passing through a point 25 mm and not more than 88 mm cranial to the anterior edge of the aitch bone. The hind foot is retained. Alternative specifications shall be as agreed between buyer and seller.

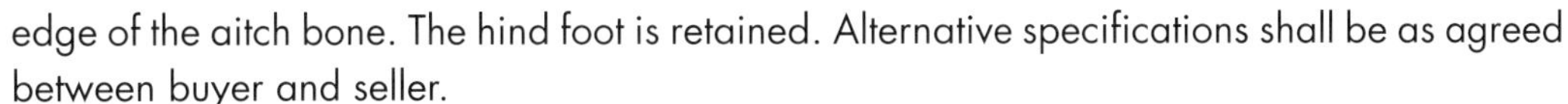

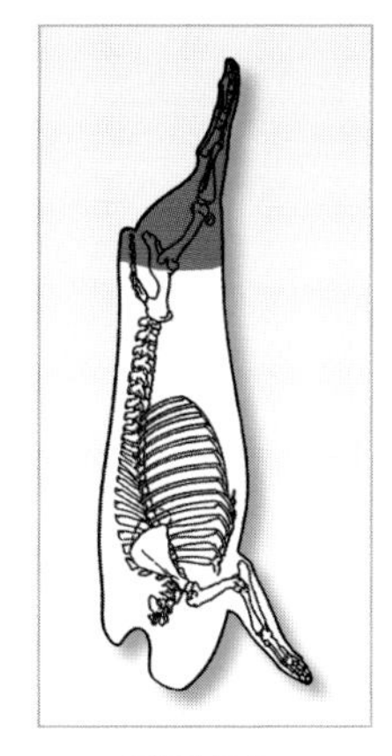

ITEM NO.
4016

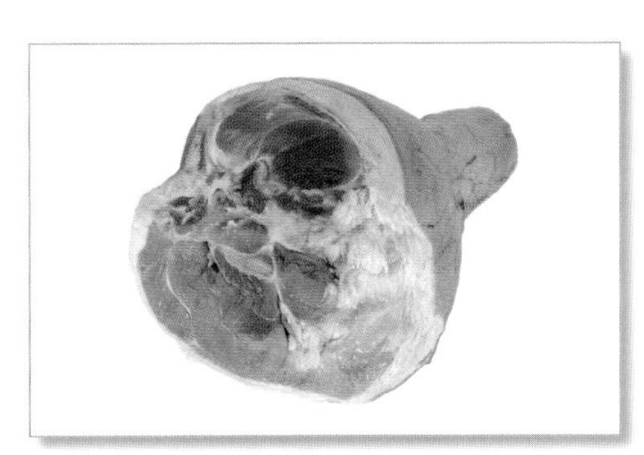

4017 LEG SHORT CUT

Style 2

Leg short cut: style 2 is the same as style 1 except the hind foot shall be removed at or slightly anterior to the hock joint. The tail (caudal) vertebrae, flank muscle (M. rectus abdominis), M. cutaneus trunci and exposed lymph glands shall be removed. The skin and collar fat over the M. semimembranosus shall be smooth and well rounded such that the innermost curvature of the skin is trimmed back at least half the distance from the stifle joint to the posterior edge of the aitch bone. The skin overlying the medial side (inside) of the M. quadriceps femoris shall be removed and fat overlying the M. quadriceps femoris and pelvic area shall be removed close to the lean. The aitch bone shall be partially removed with the ischium left intact.

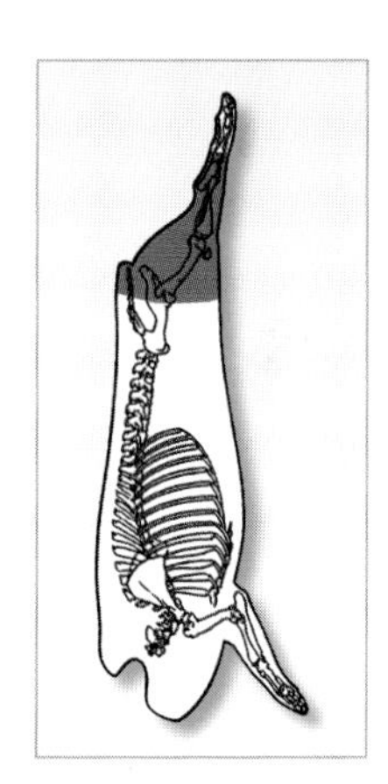

ITEM NO.
4017
4018

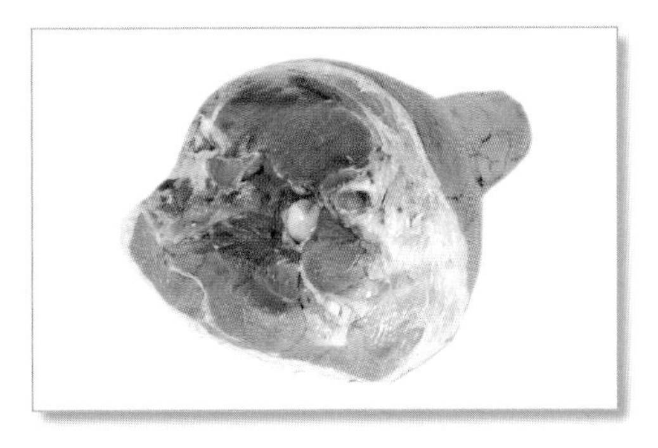

4018 LEG SHORT CUT

Style 3

Leg short cut: style 3 is the same as style 2 except the ischium, vertebrae, aitch bone, and overlying lean and fat are removed. The foot is removed at or slightly anterior to the hock joint, by a cut half the distance between the hock and stifle joints, or at other designated locations.

To be specified:

- Skin removed

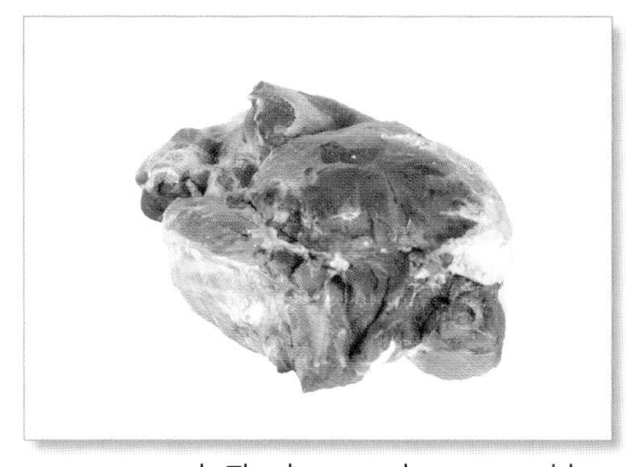

4200 LEG LONG CUT (BONELESS)

Leg long cut (boneless) is separated from the split carcase side (item 4001) by a straight cut approximately perpendicular to a line parallel to the vertebral column between the 6th and 7th lumbar vertebrae and passing through a point immediately anterior to the hip bone (ilium) and related cartilage. All bones and cartilage shall be removed. The flank and associated flank fat are removed. The leg can be seamed boned or tunnel boned. Skin shall be removed.

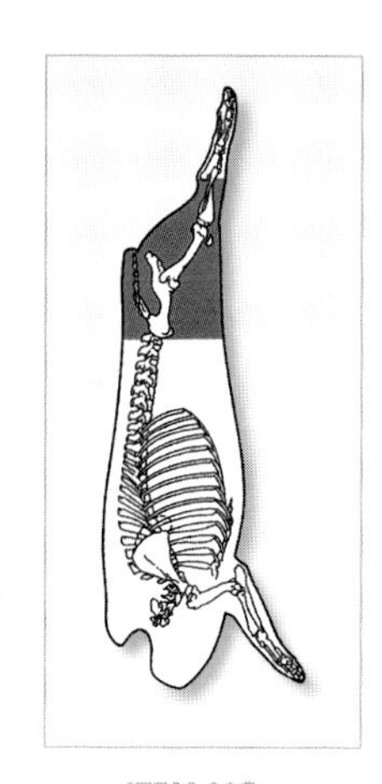

ITEM NO.
4200

To be specified:

- Skin retained
- Tunnel boned
- Seam boned

4300 OUTSIDE*

Outside shall consist of the outside muscles from the leg (M. biceps femoris and M. semitendinosus). The inner shank may remain; the M. flexor digitorum superficialis and associated fat must be removed. All external skin is removed.

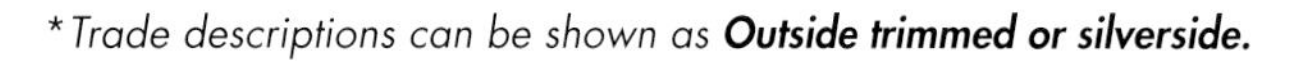
*Trade descriptions can be shown as **Outside trimmed or silverside.***

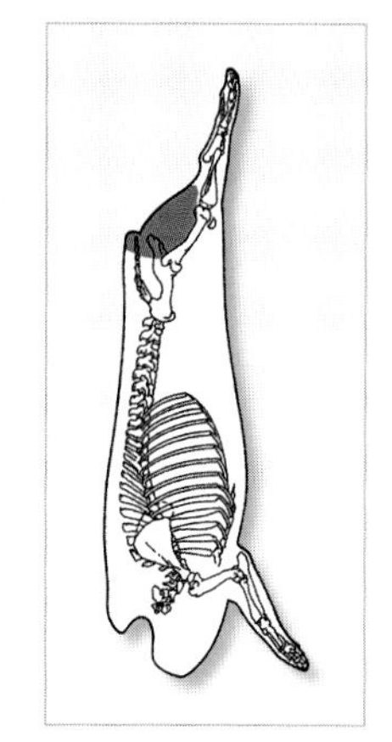

ITEM NO.
4300

4301 OUTSIDE EYE

Outside eye is prepared from an outside portion of the pork leg. It shall consist of the M. semitendinosus only.

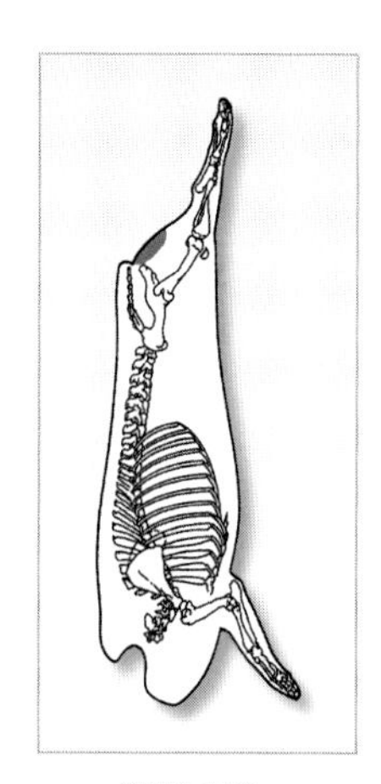

ITEM NO.
4301

4290 INSIDE

Inside shall consist of the M. semimembranosus and related muscles of the inside portion of the leg which are removed from the outside and knuckle (tip) portions of the leg along the natural seam. All bones, cartilage and heavy connective tissue shall be removed.

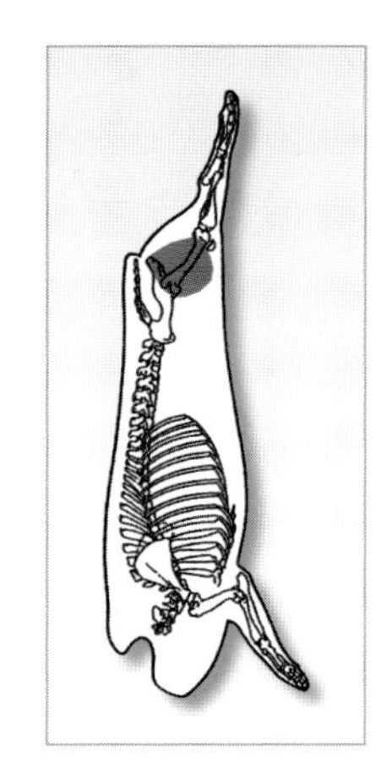

ITEM NO.
4290

4310 KNUCKLE (TIP)

Knuckle (tip) is prepared from the ventral portion of a boneless leg by removal along the seams between the knuckle and inside and knuckle and silverside. The knuckle consists of the M. rectus femoris, M. vastus medialis, M. vastus intermedius and M. vastus lateralis. The cap portion (M. tensor fascia latae) is also retained.

To be specified:

- Cap portion (M. tensor fascia latae) removed

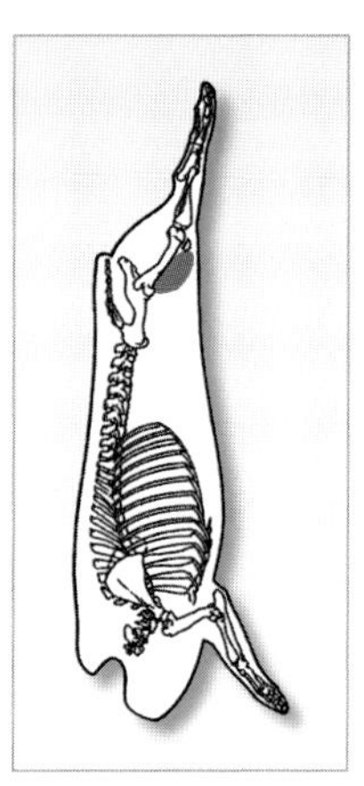

ITEM NO.
4310

4311 3-WAY LEG

(Inside – 4290, outside – 4300 and knuckle – 4310)

3-way leg is prepared from a leg short cut. It shall consist of the following primal cuts removed along the natural seams: inside (item 4290), outside (item4300), and knuckle (tip) (item 4310).

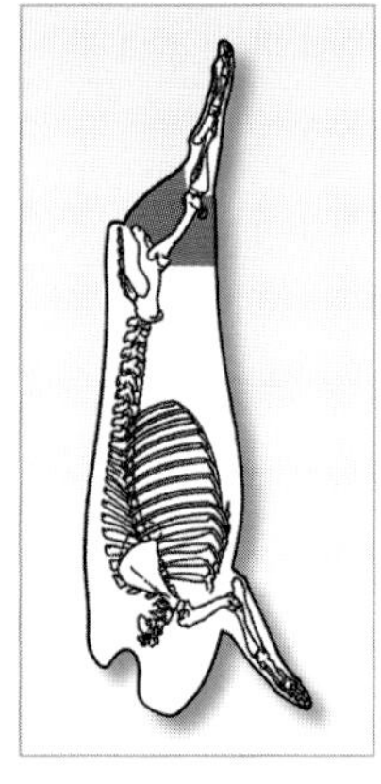

ITEM NO.
4311

4312 4-WAY LEG

(Inside – 4290, outside – 4300, inner shank (heel) and knuckle – 4310)

4-way leg is prepared from a leg short cut and consists of the following primal cuts removed along the natural seams: inside (item 4290), outside (item 4300), inner shank (heel) and knuckle (tip) (item 4310).

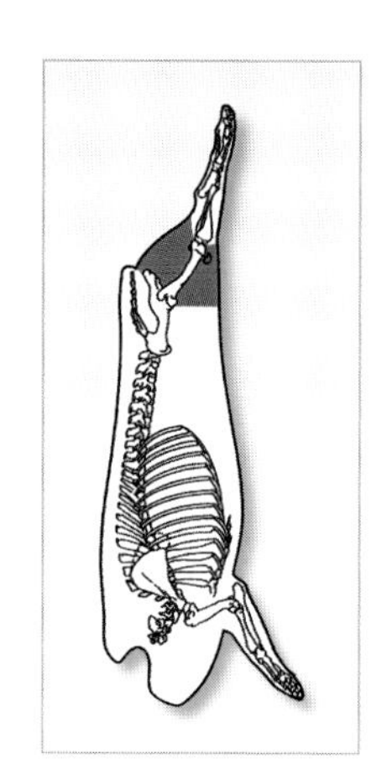

ITEM NO.
4312
4313

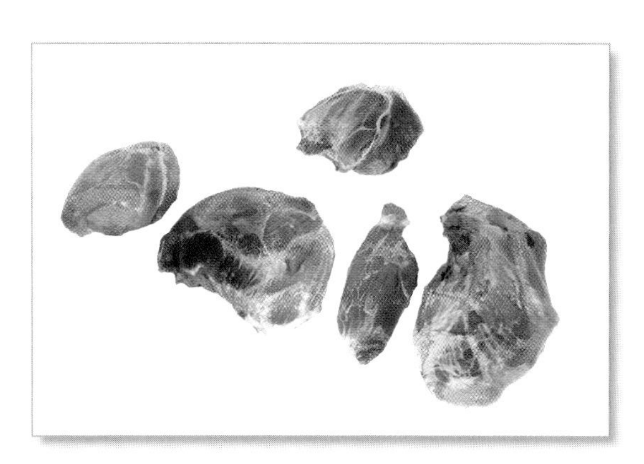

4313 5-WAY LEG

(Inside – 4290, outside eye – 4301, outside flat, inner shank (heel) and knuckle (tip) – 4310)

5-way leg is prepared from a leg short cut and consists of the following primal cuts removed along the natural seams: inside (item 4290), outside eye (item 4301), outside flat, inner shank (heel) and knuckle (tip) (item 4310).

4314 6-WAY LEG

(Inside - 4290), outside eye - 4301, outside flat, inner shank (heel), rump (sirloin) - 4130) and knuckle (tip) - 4310)

6-way leg is prepared from a leg long cut and consists of the following primal cuts removed along the natural seams: inside (item 4290), outside eye (item 4301), outside flat, inner shank (heel), rump (sirloin) (item 4130) and knuckle (tip) (item 4310).

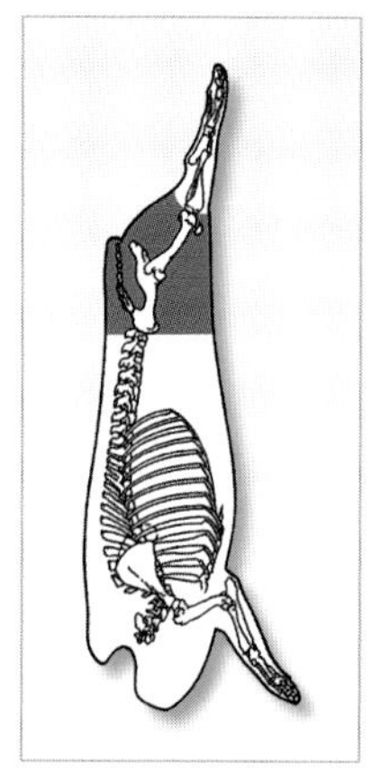

ITEM NO.
4314

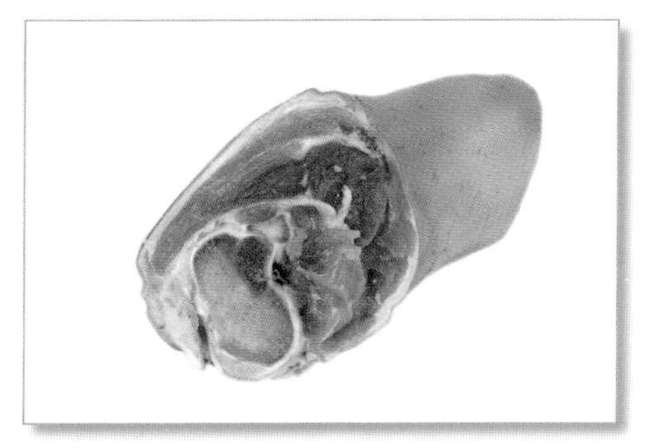

4172 HOCK LEG*

Hock leg is prepared from a leg (item 4013) by the removal of the hind foot at the tarsal joint and the leg at the stifle joint. Skin shall remain.

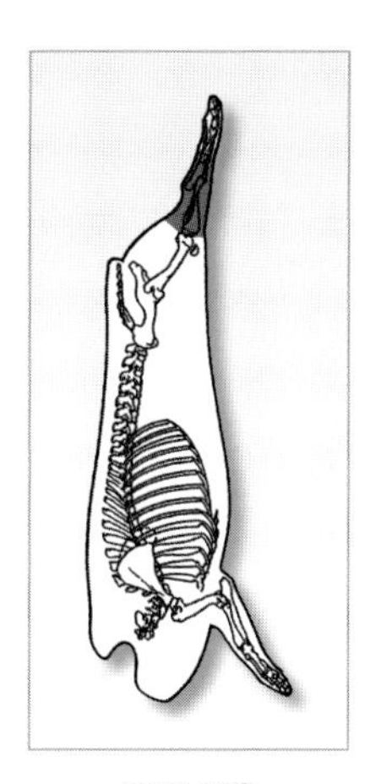

ITEM NO.
4172

To be specified:

- Skin removed
- Frenched

* *Trade descriptions can be shown as* ***Ossobucco.***

4176 HIND FEET (TROTTER)

Hind feet (trotter) are removed from a leg at the tarsal joint severing the hind foot from the leg. Skin shall remain.

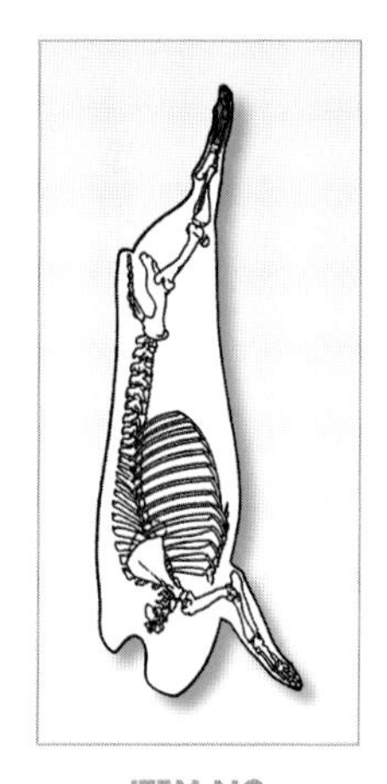

ITEM NO.
4176

To be specified:

- Skin removed

4069 - 4072 MIDDLE

Middle is derived from a split carcase side (item 4001) by removal of the leg (item 4013) and forequarter (item 4021) at the specified cutting lines. The diaphragm and tenderloin are removed.

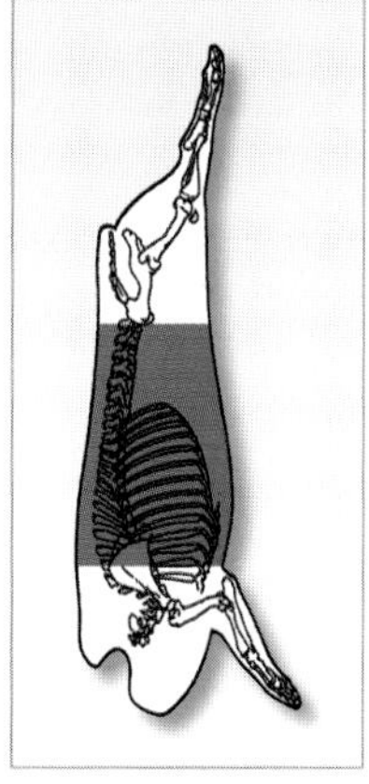

ITEM NO.
4069 (13-ribs)
4070 (12-ribs)
4071 (11-ribs)
4072 (10-ribs)

To be specified:

- Skin removed
- Diaphragm retained
- Blade bone (scapula) removed
- Tenderloin retained

4079 - 4082 BELLY

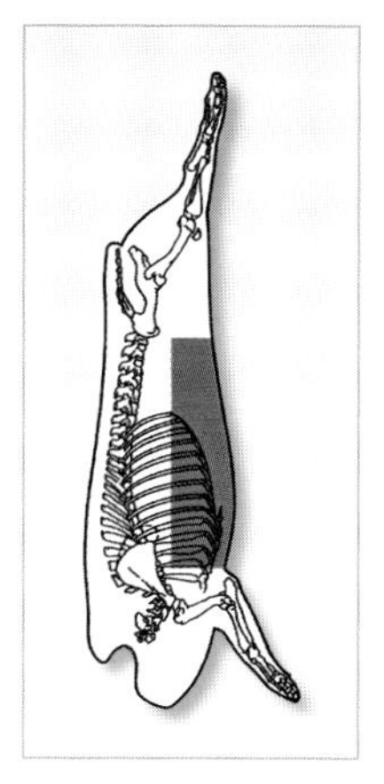

Belly (bone-in) is prepared from the middle (item 4069) by the removal of the loin-long (item 4140). The skin is retained. Quantity of leaf fat shall be removed. The anterior (shoulder) and posterior (leg) ends of the belly shall be reasonably straight and parallel. No side of the belly shall be more than 5 cm longer than its opposing side. The width of the flank muscle (M. rectus abdominis) shall be at least 25 per cent of the width of the belly on the leg end. The fat on the ventral side of the belly and adjacent to the flank shall be trimmed to within 2 cm from the lean. The belly shall be free of enlarged, soft, porous, dark, or seedy mammary tissue.

ITEM NO.
4079 (13-ribs)
4080 (12-ribs)
4081 (11-ribs)
4082 (10-ribs)

To be specified:

- Number of ribs required
- Skin removed
- Diaphragm retained
- Width of belly

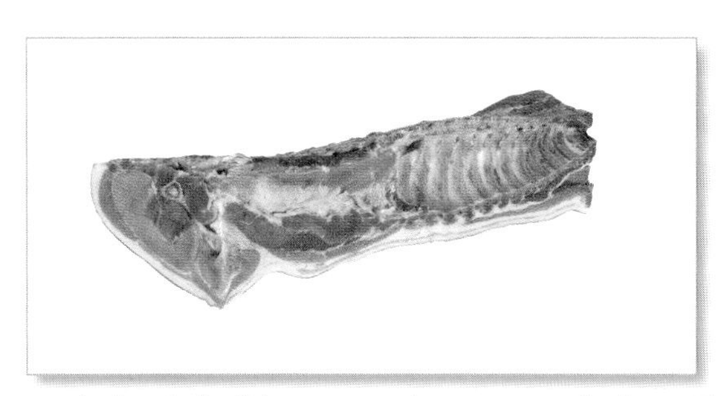

4140 - 4147 LOIN – LONG

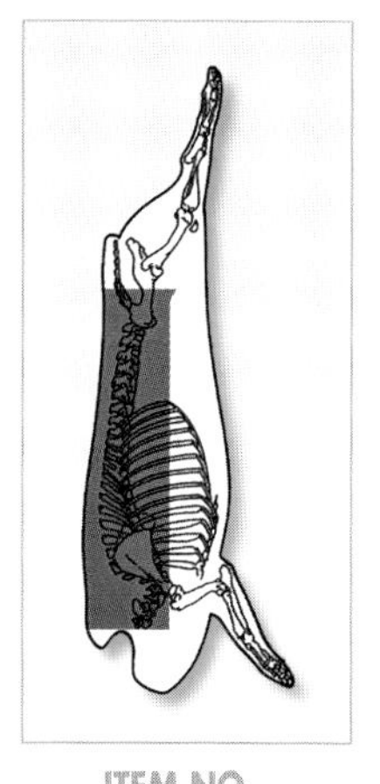

Style 1

Loin - long is the remaining dorsal portion of the carcase side after the removal of the leg short cut (item 4016) and shoulder picnic and belly (item 4335). Lumbar fat (on the inside surface covering the tenderloin) shall be trimmed to practically free. The tenderloin is retained.

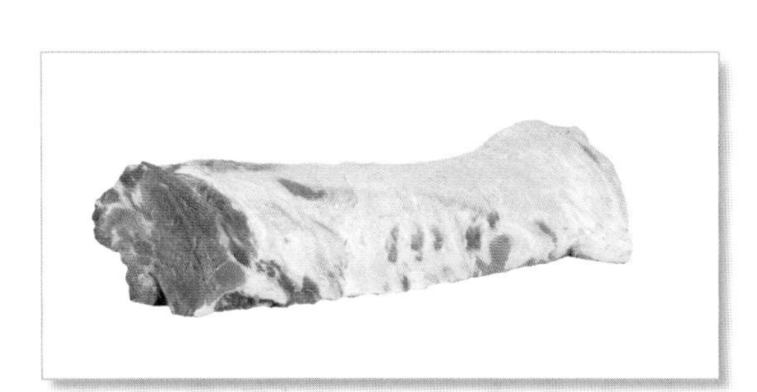

Style 2

The blade bone, related cartilage and overlying muscles (in their entirety) and fat shall be removed.

ITEM NO.
4140 (17-ribs)
4141 (16-ribs)
4142 (15-ribs)
4143 (14-ribs)
4144 (13-ribs)
4145 (12-ribs)
4146 (11-ribs)
4147 (10-ribs)

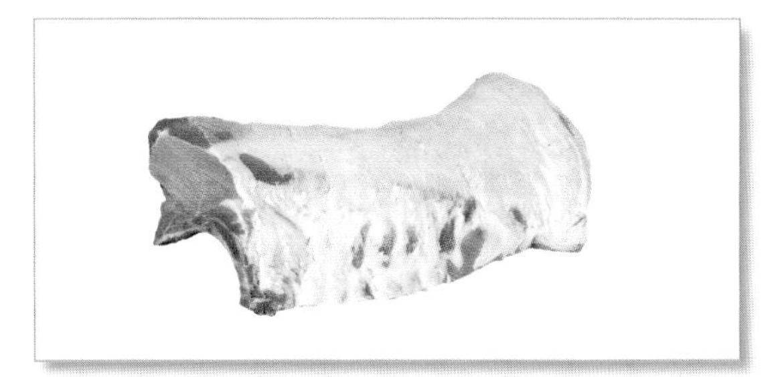

Style 3

The blade portion shall be removed to leave not more than eight ribs present and the M. longissimus dorsi shall be at least twice as large as the M. spinalis dorsi.

To be specified:

- Skin removed
- Level of fat trim

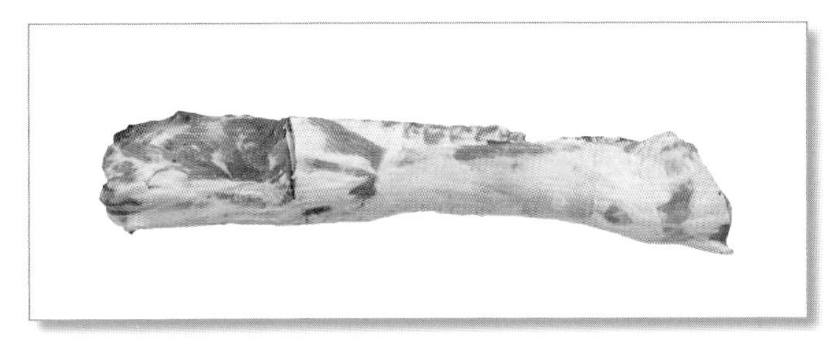

4108 - 4111 LOIN – LONG (BLADE REMOVED)

Style 1

Loin – long (blade removed) is prepared from the loin-long (item 4140). The skin is removed. The loin is further prepared by removal of the chine bone. Feather bones and ribs shall be retained.

Style 2

The blade portion shall be removed to leave not more than eight ribs present and the M. longissimus dorsi shall be at least twice as large as the M. spinalis dorsi.

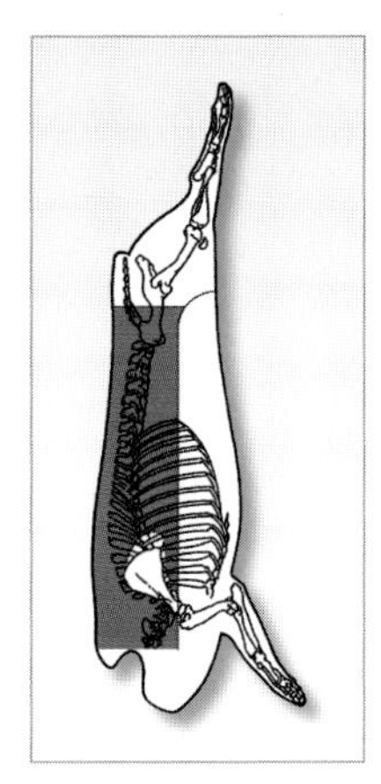

ITEM NO.

4108 (13-ribs or more)
4109 (12-ribs)
4110 (11-ribs)
4111 (10-ribs)

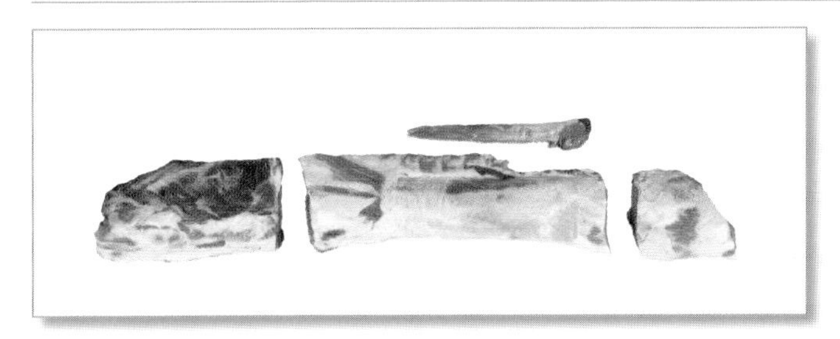

4113 LOIN – LONG (4-WAY)

Loin – long (4-way) is prepared from a loin-long (item 4108), skin removed. The loin is cut into four distinct portions: loin - centre cut (item 4101) removed at the specified rib, tenderloin (item 4280), sirloin (rump) (item 4130) and shoulder inside (item 4046) removed at the specified rib.

To be specified:

- Rib cutting line for loin and shoulder removal points

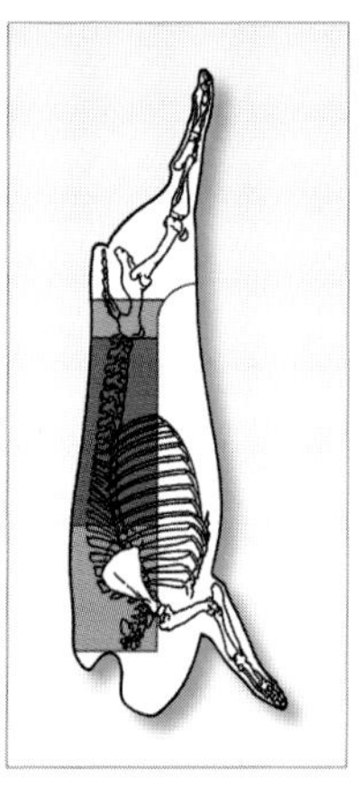

ITEM NO.

4113

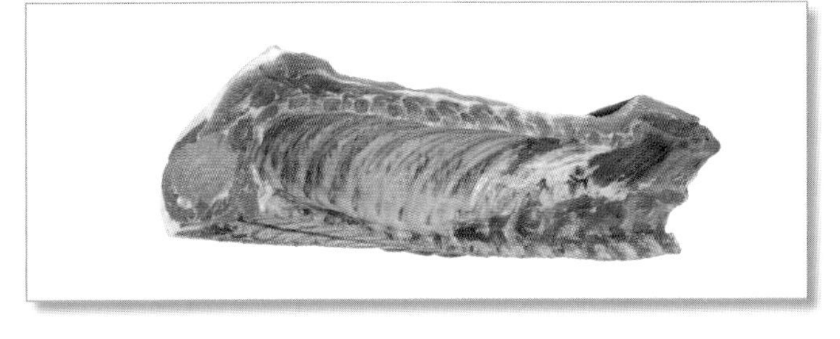

4098 - 4101 LOIN – CENTRE CUT

Loin - centre cut is prepared from the middle (item 4069) by the removal of the belly by a cut at a specified distance from the ventral edge of the eye muscle and parallel to the backbone (measured from the cranial end). Skin, blade (scapula) bone and associated cartilage shall be removed unless otherwise specified.

To be specified:

- Ventral cutting line (tail length)
- Diaphragm retained
- Tenderloin retained
- Chine bone removed

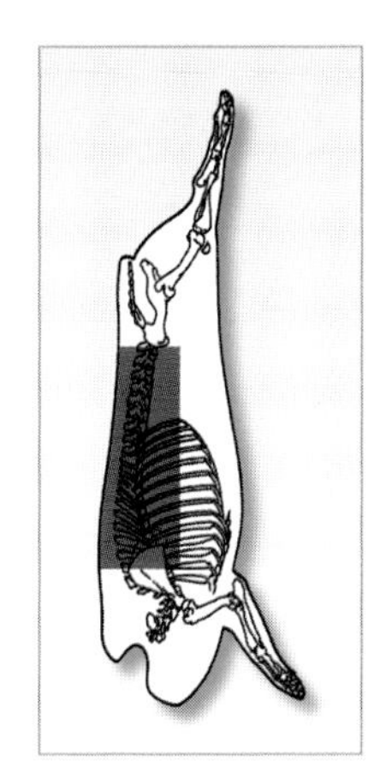

ITEM NO.

4098 (13-ribs)
4099 (12-ribs)
4100 (11-ribs)
4101 (10-ribs)

4102 - 4105 SEMIBONELESS LOIN – CENTRE CUT

Semiboneless loin – centre cut is prepared from the loin-centre cut by removing the chine bone; the feather bones and ribs shall remain.

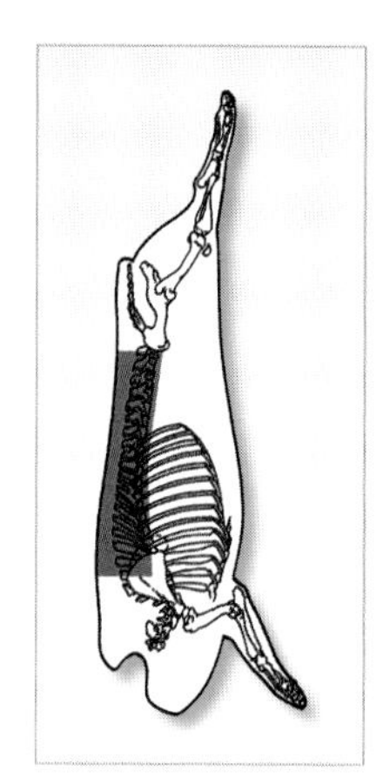

ITEM NO.
4102 (13-ribs)
4103 (12-ribs)
4104 (11-ribs)
4105 (10-ribs)

4130 SIRLOIN (RUMP)

Sirloin (rump) is prepared from a leg long cut (item 4013). The sirloin is removed by a cut across the leg at a specified distance from the acetabulum. Unless specified a portion of the tenderloin may be retained.

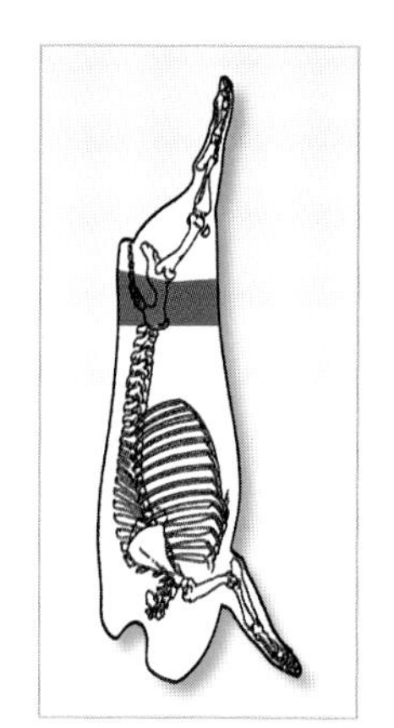

To be specified:

- Skin removed
- Removal point from loin-long (item 4140)

ITEM NO.
4130

4159 LOIN RIBLETS

Loin riblets are derived from the transverse processes and associated lean from the lumbar vertebrae of any bone-in pork loin after removal of the tenderloin and the loin eye. Loin riblets shall contain no less than four transverse processes (paddle/finger bones), be held intact by associated lean and include no more than two rib bones. This item shall be trimmed practically free of surface fat.

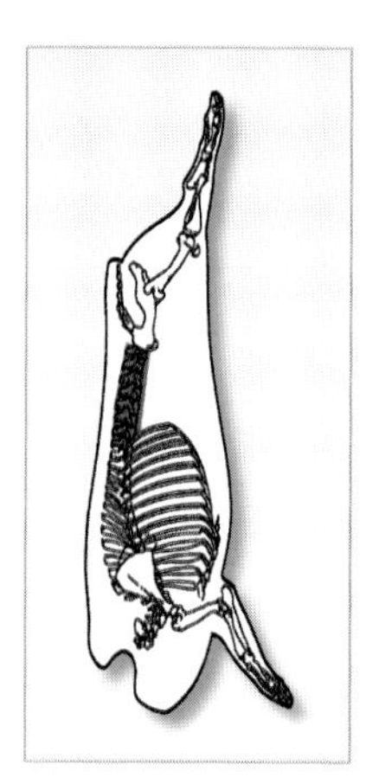

ITEM NO.
4159

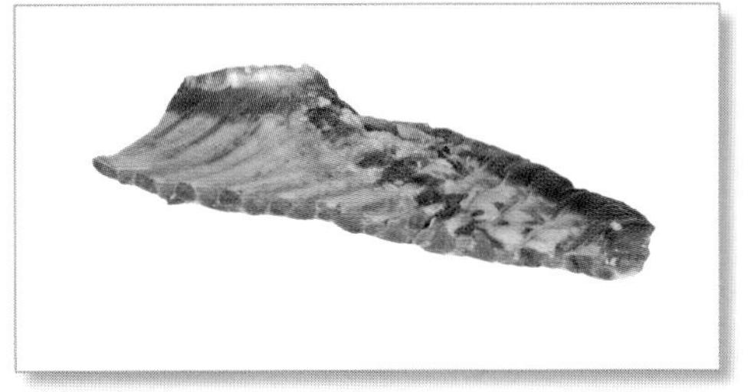

4160 BELLY RIBS*

Belly ribs are prepared from a belly bone-in by removal of the ribs, costal cartilages and intercostal muscles and shall consist of at least eight ribs.

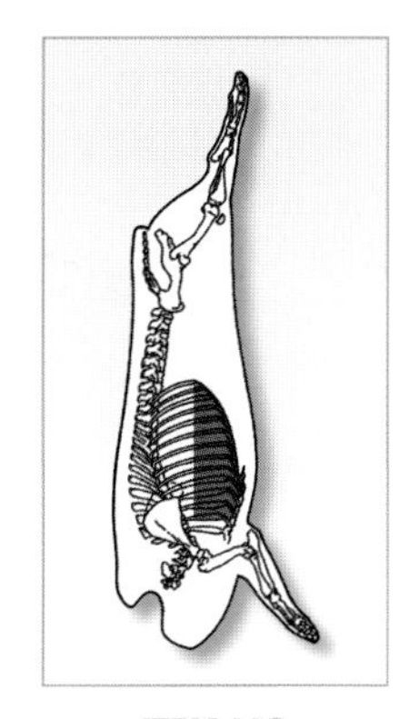

To be specified:

- Number of ribs required
- Diaphragm retained
- Width of belly ribs
- Sternum removed

Trade descriptions can be shown as **Spare ribs.*

ITEM NO.
4160

4161 BACK RIBS*

Back ribs are prepared from a bone-in loin by the removal of all bones and cartilage and shall consist of at least eight ribs and related intercostal meat. The back ribs section shall be intact, and the bodies of the thoracic vertebrae shall be removed except that small portions of the vertebrae may remain between the rib ends.

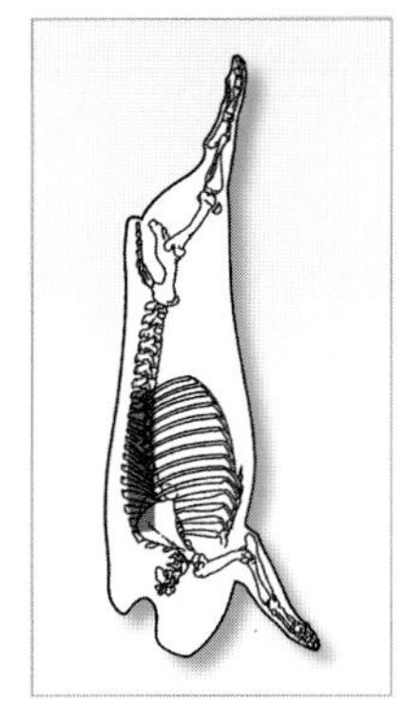

To be specified:

- Number of ribs required
- Diaphragm retained
- Width of back ribs
- Peritoneum removed from the inside surface of the ribs and intercostals muscles

Trade descriptions can be shown as **Loin ribs.*

ITEM NO.
4161

4162 FULL RIB PLATE

Full rib plate is prepared from the middle by complete removal of the entire rib plate in one piece and the attached intercostal muscles. The diaphragm is removed.

To be specified:

- Number of ribs required
- Diaphragm retained
- Costal cartilage removed

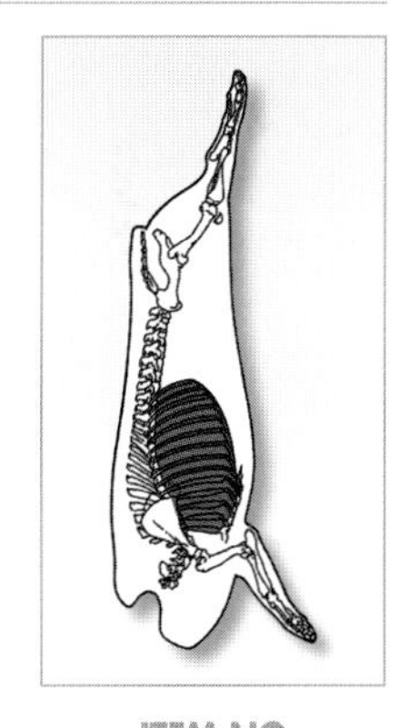

ITEM NO.
4162

4163 ST. LOUIS STYLE RIBS

St. Louis style ribs are prepared from belly ribs by removal of the sternum and ventral portion of the costal cartilage. At least eight ribs remain.

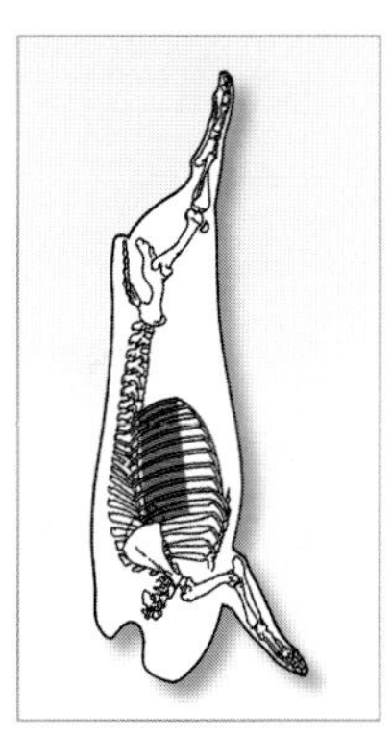

ITEM NO.
4163

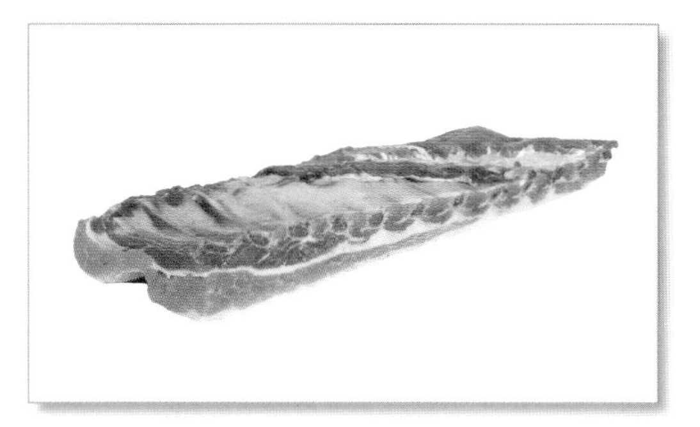

4164 SHORT RIBS

Short ribs will be removed from the dorsal side of the belly by a straight cut along the length of the belly. The ribs will consist of a width of approximately 120 mm.

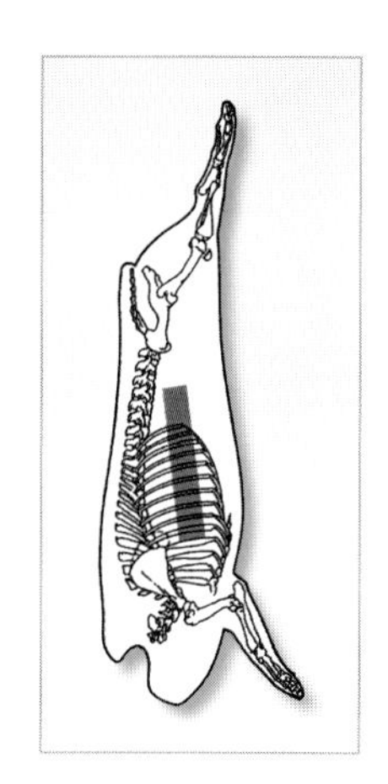

ITEM NO.
4164

4319 - 4322 MIDDLE

Middle is derived from a carcase side (item 4001) by the removal of the leg (item 4013) at the specified lumbar vertebrae and forequarter (item 4021) at the specified rib or thoracic vertebrae. The diaphragm and tenderloin are removed. All bones and cartilage are removed.

To be specified:

- Skin removed
- Diaphragm retained
- Rib bones sheet/string boned

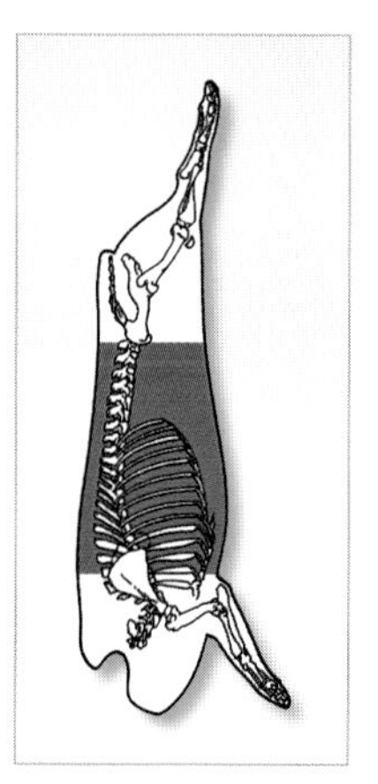

ITEM NO.
4319 (13-ribs)
4320 (12-ribs)
4321 (11-ribs)
4322 (10-ribs)

4340 - 4343 LOIN

Loin is prepared from the middle (item 4319) by the removal of the boneless belly (item 4329) by a cut at a specified distance from the ventral edge of the eye muscle and parallel to the backbone (measured from the cranial end).

To be specified:

- Skin retained
- Ventral cutting line (tail length)

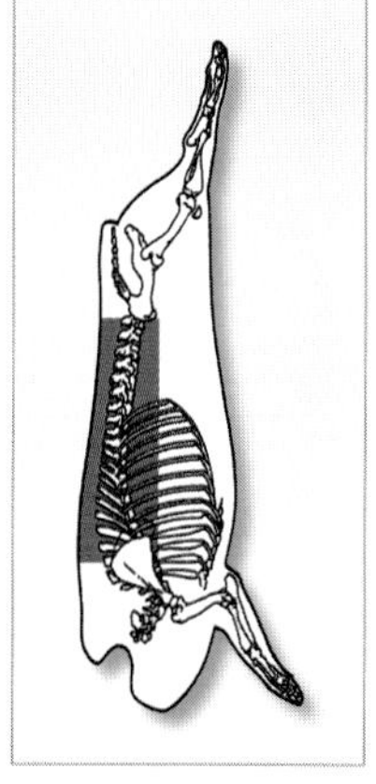

ITEM NO.
4340 (13-ribs)
4341 (12-ribs)
4342 (11-ribs)
4343 (10-ribs)

4361 EYE OF LOIN

Eye of loin is prepared from a loin (item 4340) and consists of the eye muscle portion (M. longissimus dorsi) removed along the natural seam. Intercostals and attached muscle portions are removed.

To be specified:

- Denuded of all fat
- Silverskin removed
- M. multifidus dorsis retained

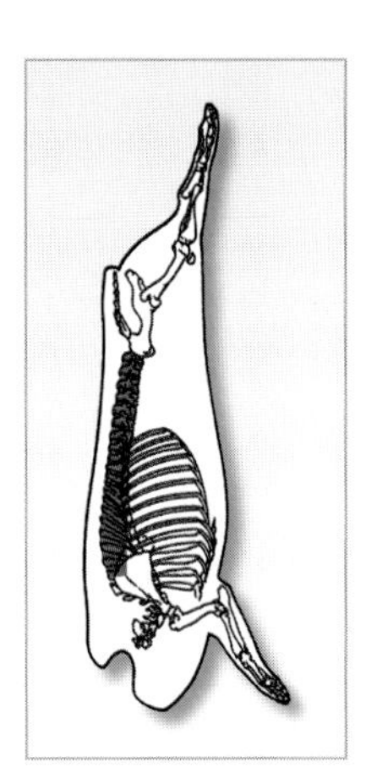

ITEM NO.
4361

4360 EYE OF SHORTLOIN

Eye of shortloin is prepared from a loin (item 4340) and consists of the eye muscle portion (M. longissimus dorsi) located from the tenth thoracic vertebrae to the junction of lumbar sacral vertebrae and is carefully removed along the natural seam. Intercostals and attached muscle portions are removed.

To be specified:

- Denuded of all fat
- Silverskin removed
- M. multifidus dorsis retained

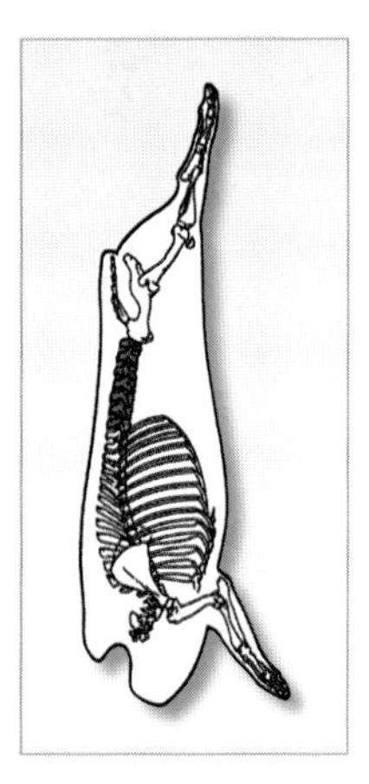

ITEM NO.
4360

4305 SIRLOIN (RUMP) BONELESS

Sirloin (rump) boneless is prepared from a sirloin (rump) bone-in (item 4130) by removing all bone, cartilage, surface fat and remaining tenderloin portion.

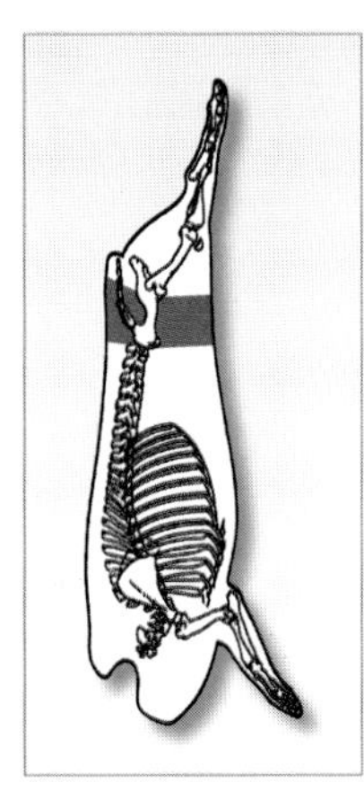

ITEM NO.
4305

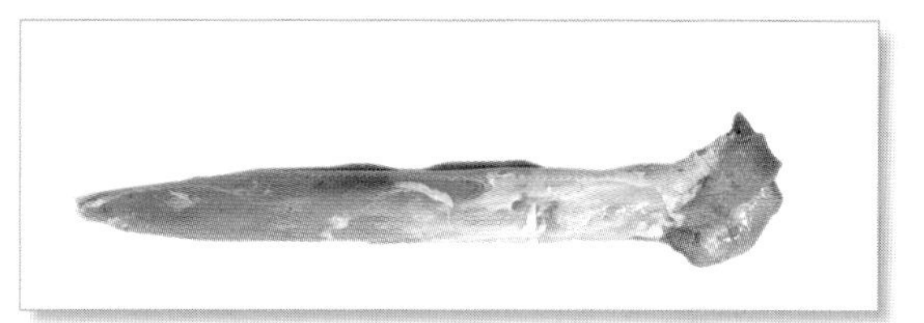

4280 TENDERLOIN

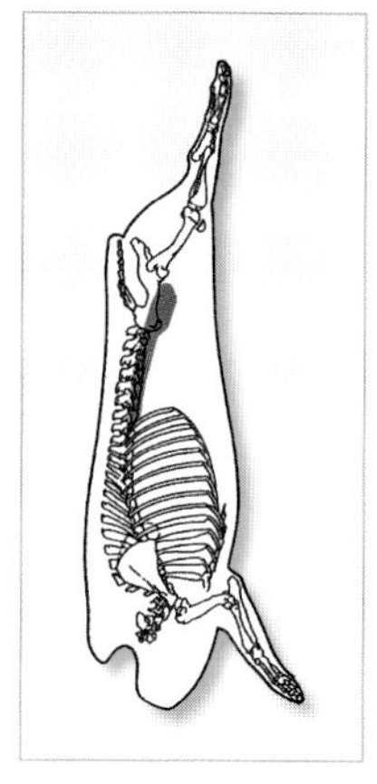

Tenderloin consists of the M. psoas major and M. iliacus, which are along the ventral surface of the lumbar vertebrae and lateral surface of the ilium. The side strap muscle (M. psoas minor) is removed. The tenderloin shall be practically free of surface fat.

To be specified:

- Side strap (M. psoas minor) retained
- Head muscle (M. iliacus) removed

ITEM NO.
4280

4329 - 4332 BELLY (BONELESS)

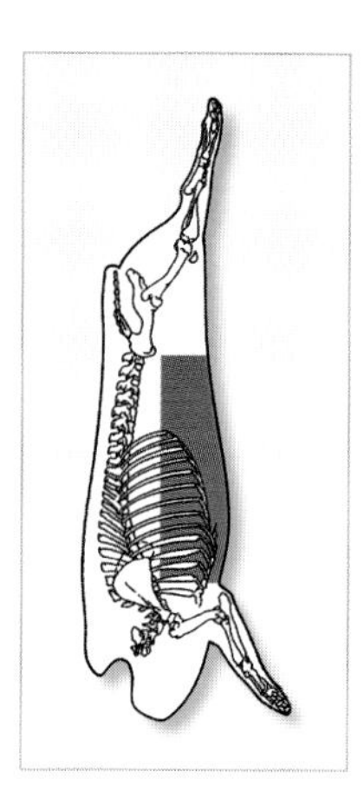

Belly is prepared from the belly bone-in (item 4079) by removal of the ribs.

Style 1
Ribs are individually removed from the belly leaving the costal cartilage, diaphragm and finger meat (intercostal meat) intact and firmly attached to the belly.

ITEM NO.
4329 (13-ribs)
4330 (12-ribs)
4331 (11-ribs)
4332 (10-ribs)

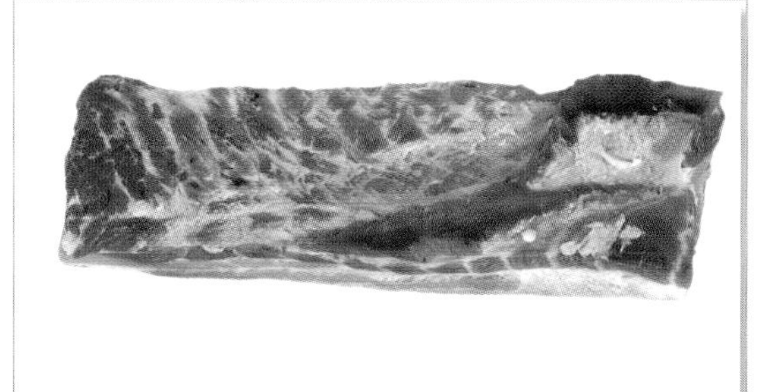

Style 2
Ribs are removed intact from the belly along with the costal cartilage, diaphragm and finger meat (intercostal meat). The belly shall be free of scores and "snowballs" (exposed areas of fat) which measure 50 cm^2 or more.

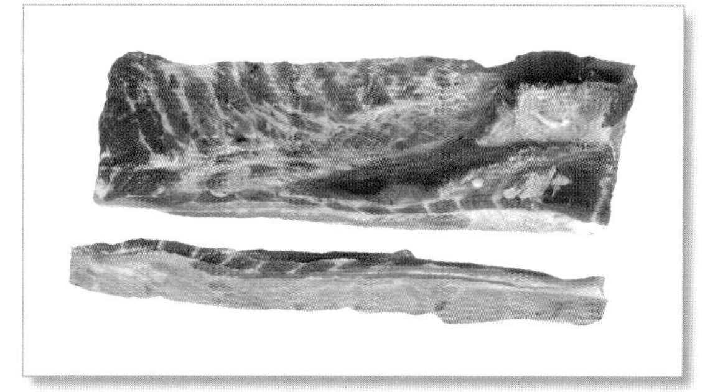

Style 3
Same as style 2 except that the teat line shall be removed by a straight cut.

To be specified:

- Number of ribs required
- Skin removed
- Diaphragm retained
- Width of belly

4333 BELLY (FLANK ON)

Belly (flank on) is the remaining primal after the shoulder-picnic (item 4046) has been removed from the shoulder-picnic and belly (item 4335) along the specified rib.

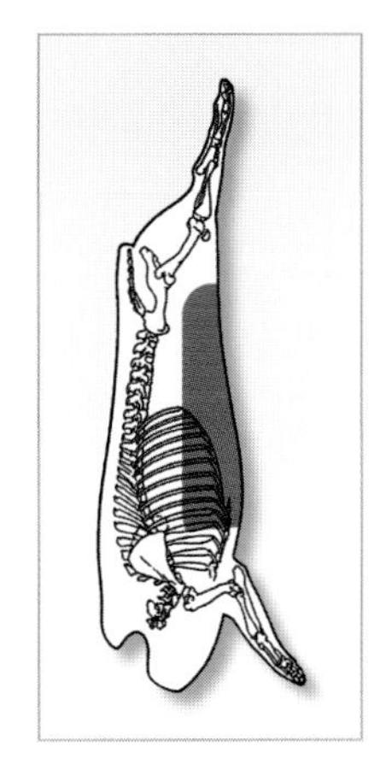

To be specified:

- Flank removed
- Belly ribs removed (item 4160)
- Skin removed
- Belly edge removed

ITEM NO.
4333

4335 SHOULDER-PICNIC AND BELLY

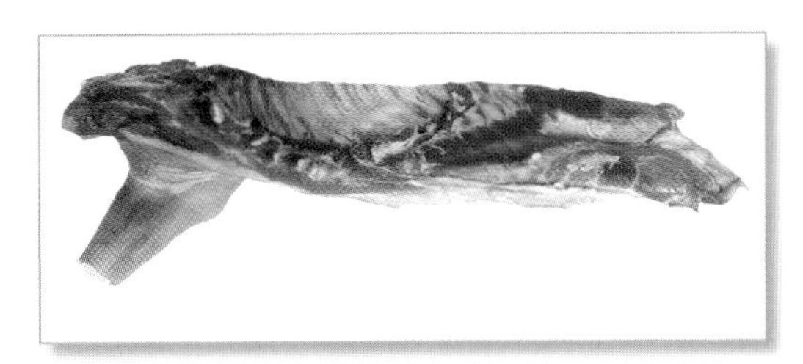

Shoulder-picnic and belly is prepared from a carcase side. The shoulder-picnic and belly separation point is made by a cut commencing at the cranial end and at a specified distance from the vertebral column through the joint of the blade and humerus bones and cut parallel to the chine edge for the full length of the loin to the tip of and including the extended muscles of the flank.

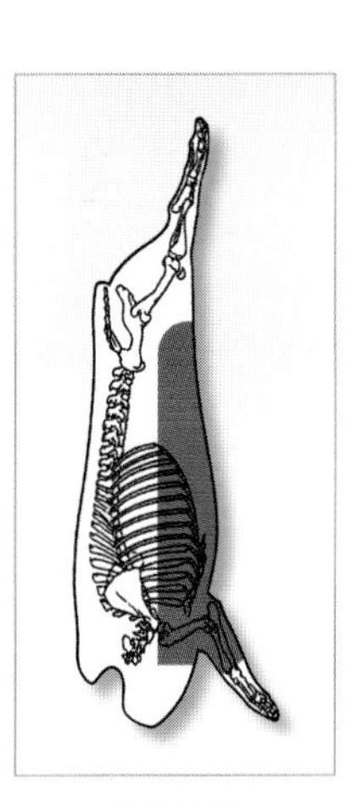

ITEM NO.
4335

4029 - 4032 SHOULDER – SQUARE CUT

Shoulder-square cut is separated from the carcase side (item 4001) by a straight cut, approximately perpendicular to the length of the carcase side at the specified rib. The head, jowl and breast flap shall be removed by a straight cut approximately parallel with the loin side which is anterior to, but not more than 25 mm from the innermost curvature of the ear dip. The foot and breast flap is removed.

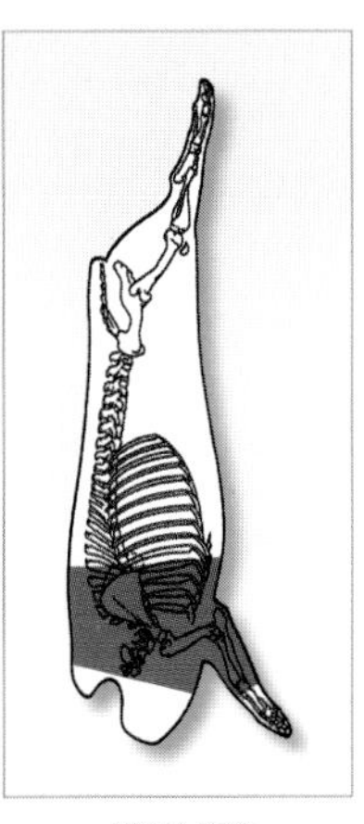

To be specified:

- Number of ribs
- Skin removed
- Foot (trotter) retained
- Neck bones, ribs, breast bones and associated cartilage removed

ITEM NO.
4029 (1-rib)
4030 (2-ribs)
4031 (3-ribs)
4032 (4-ribs)

4044 SHOULDER OUTSIDE

Shoulder outside is prepared from the shoulder-square cut (item 4029) and consists of the scapular, humerus and foreshank bones together with associated muscles. The shoulder outside is separated from the shoulder-square cut by a cut starting under (medial) the front leg, passing through the M. pectoralis superficialis, the natural seam between the M. serratus ventralis and the M. lattissimus dorsi, the natural seam between the M. serratus ventralis and the M. subscapularis, the natural seam between the M. serratus ventralis and the medial side of the scapula to a point immediately dorsal to the cartilage of the scapula. All sides shall be trimmed following the natural curvature of the major muscles and the scapula. The posterior side shall not expose the M. triceps brachii. The skin and foot is retained.

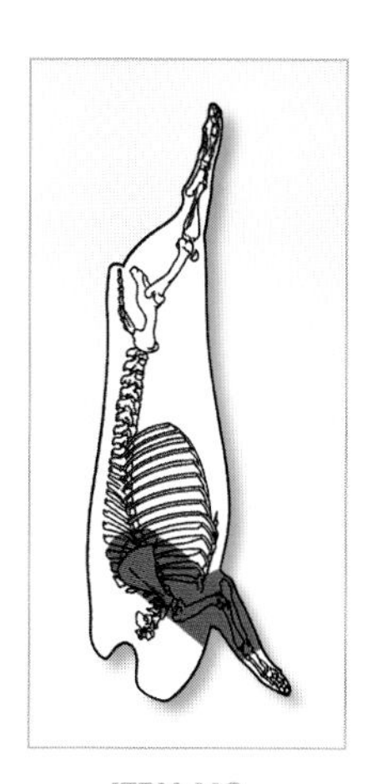

ITEM NO.
4044

To be specified:

- Foot (trotter) removed
- Breast flap retained
- Skin removed
- Ribs removed

4045 SHOULDER OUTSIDE (3-WAY)

Shoulder outside (3-way) is prepared from the forequarter (item 4005) removed from carcase side - block ready (4-way) (item 4004) which has the vertebrae and associated meat of the neck and forequarter removed. The shoulder ribs (item 4164) are removed. The remaining portion, shoulder outside (item 4044) is cut into two pieces by a separating cut (cranial to caudal) running through the joint of the blade and humerus bones.

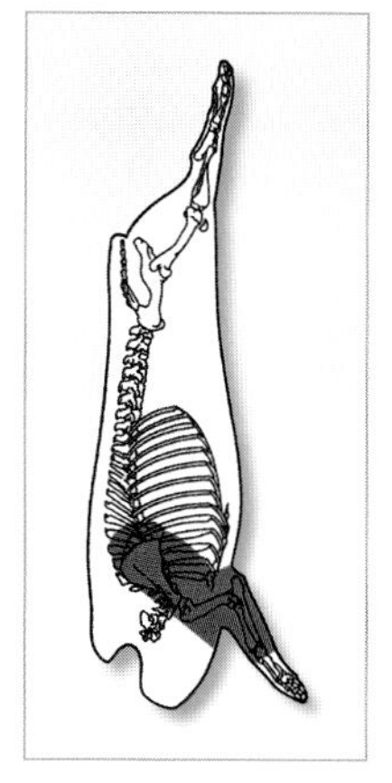

ITEM NO.
4045

4050 - 4055 SHOULDER LOWER HALF*

Shoulder lower half is prepared from the shoulder-square cut (item 4029). The shoulder lower half is separated from the upper half by a straight cut, dorsal to the shoulder joint, approximately perpendicular to the loin/shoulder separation. Neck bones, ribs, breast bones, associated cartilage and breast flap (through the major crease) shall be removed. Fat and skin shall be bevelled to meet the lean on the dorsal edge.

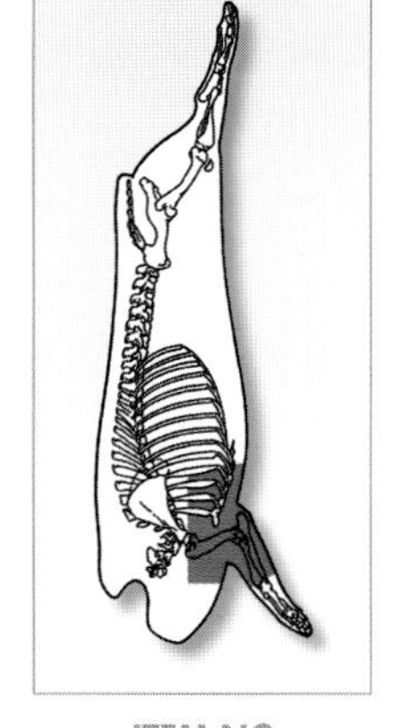

To be specified:

- Foot (trotter) removed
- Skin removed

ITEM NO.

4050 (6-ribs) 4051 (5-ribs)
4052 (4-ribs) 4053 (3-ribs)
4054 (2-ribs) 4055 (1-rib)

Trade descriptions can be shown as* ***shoulder-picnic.*

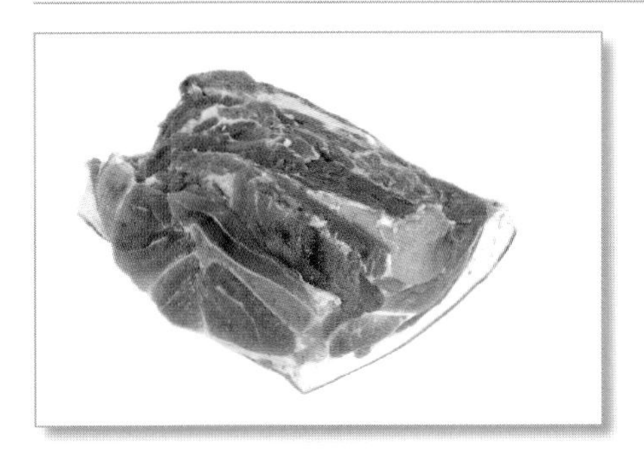

4059 - 4062 SHOULDER UPPER HALF*

Shoulder upper half is prepared from the shoulder-square cut (item 4029). The shoulder upper half is separated from the lower half by a straight cut, approximately perpendicular to the loin and shoulder separation. Fat and skin are bevelled to meet the lean on the dorsal edge. All bones and cartilage are removed. Skin is retained unless otherwise specified.

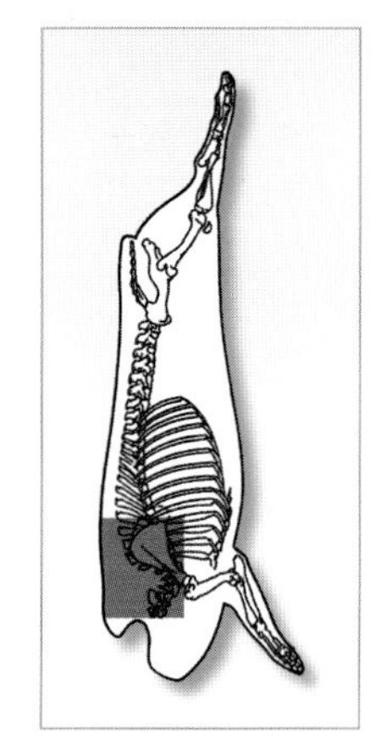

To be specified:

- Skin removed
- Scapula and related cartilage removed

ITEM NO.

4059 (4-ribs)
4060 (3-ribs)
4061 (2-ribs)
4062 (1-rib)

Trade descriptions can be shown as* ***butt or collar butt.*

4046 - 4049/4063 SHOULDER INSIDE

Shoulder outside is prepared from loin-long (blade removed) (item 4108) and is the cranial forequarter portion of the loin and removed along the specified rib.

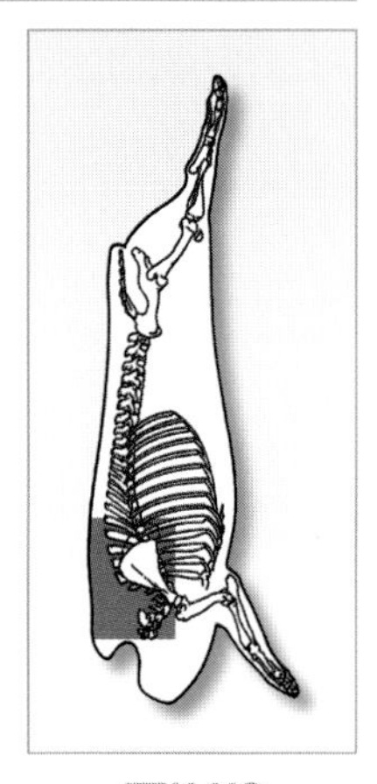

To be specified:

- Number of ribs
- Breast removal and distance from vertebrae
- Fat removed

ITEM NO.

4046 (4-ribs) 4048 (2-ribs)
4047 (3-ribs) 4049 (1-rib)
4063 (5-ribs)

4245 SHOULDER UPPER HALF (BONELESS)*

Shoulder upper half (boneless) is prepared from a shoulder upper half bone in (item 4059) by the removal of all bones, cartilage and skin. The collar butt is the dorsal portion remaining after the shoulder lower half has been removed.

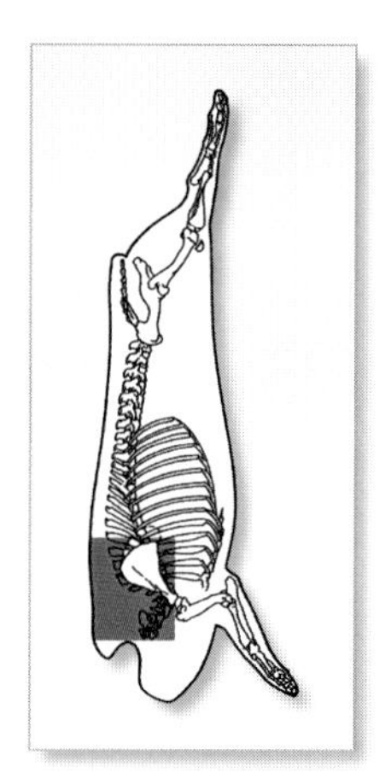

To be specified:

- Length of tail from eye of meat

ITEM NO.
4245

** Trade descriptions can be shown as **butt or collar butt.***

4241 SHOULDER INSIDE (BONELESS)

Shoulder inside (boneless) is prepared from a shoulder inside (item 4046) by removing all bone, cartilage and surface fat.

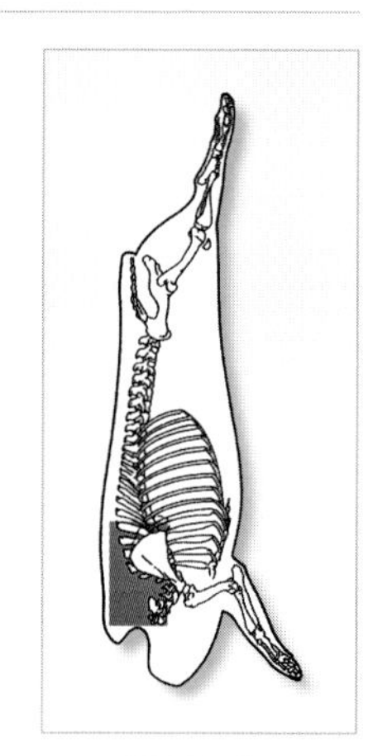

ITEM NO.
4241

4240 COLLAR BUTT – SPECIAL TRIM*

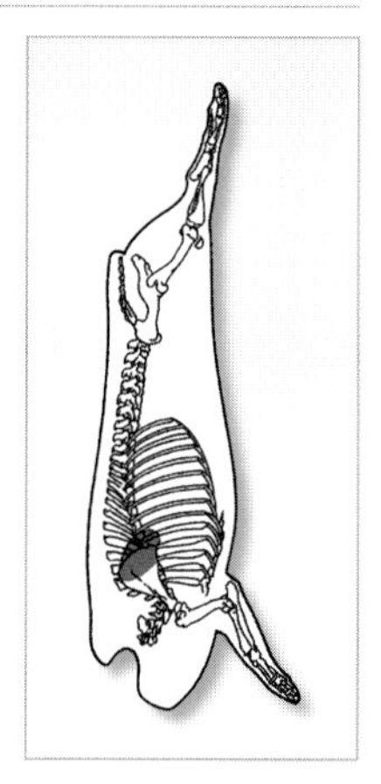

Collar butt or butt – special trim is prepared from shoulder upper half bone-in (item 4059) by the removal of the ribs, thoracic, cervical vertebrae and the shoulder lower half. The collar butt is the dorsal portion remaining after the shoulder lower half has been removed. All bone and cartilage is removed. The skin is removed from the collar butt surface. A strip of fat is retained on the lateral surface of the cut running parallel to the dorsal edge the length of the collar butt. Specify the width and thickness of the strip of fat to be retained.

ITEM NO.
4240

To be specified:

- Fat cover requirements
- Fat trim level
- Length of tail distance from eye of meat

** Trade descriptions can be shown as **butt or collar butt – special trim.***

4180 SHOULDER (M. PECTORALIS)

Shoulder (M. pectoralis) consists of the M. pectoralis profundus and M. pectoralis superficialis muscles from the breast portion of the pork shoulder. It is exposed during separation of the inside from the outside portion of the shoulder and also is located on the medial side of the lower shoulder item.

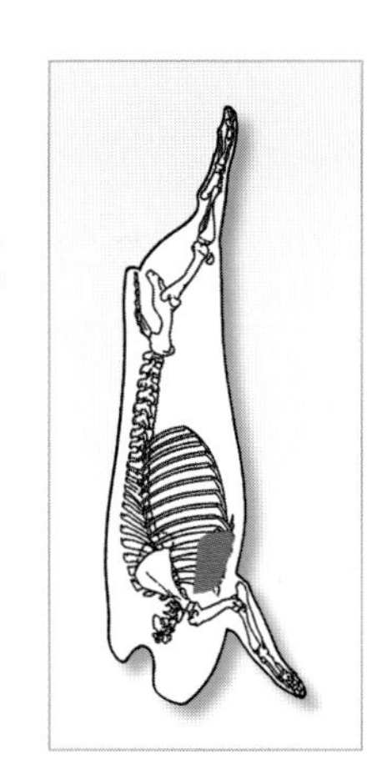

4181 SHOULDER (M. TERES MAJOR)

Shoulder (M. teres major) is removed from the medial side of the outside shoulder. It is located immediately ventral to the blade bone. It is removed by cutting along the natural seams from the adjacent muscles.

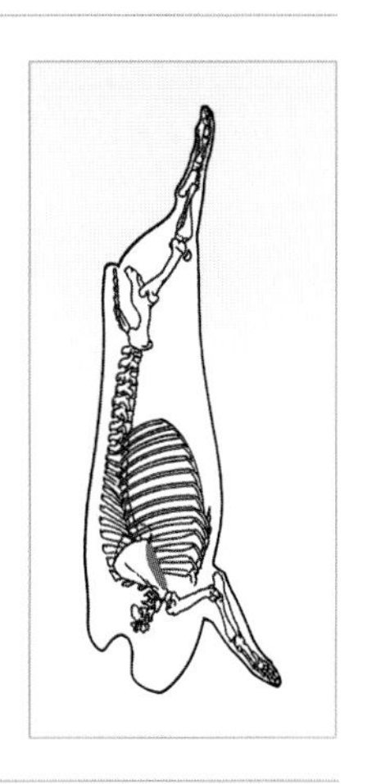

4182 SHOULDER (M. SERRATUS VENTRALIS)

Shoulder (M. serratus ventralis) consists of the M. serratus ventralis muscle from the upper shoulder and the inside shoulder. It is removed from adjacent muscles by cutting through the natural seams.

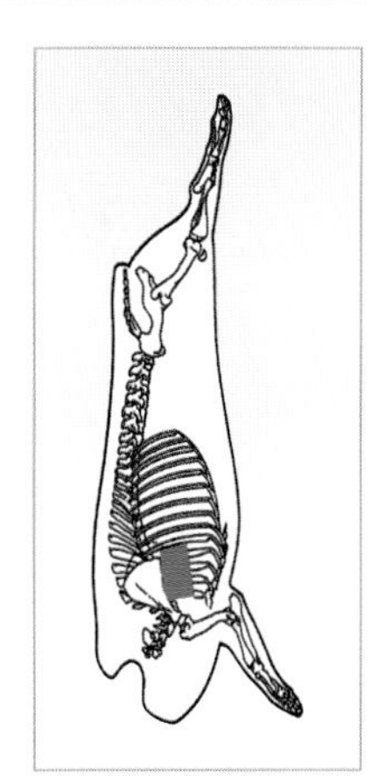

4183 SHOULDER (CUSHION)

Shoulder (cushion) consists of the M. triceps brachii muscles from the shoulder lower half and shall be practically free of fat. Tendons shall be trimmed flush with the lean.

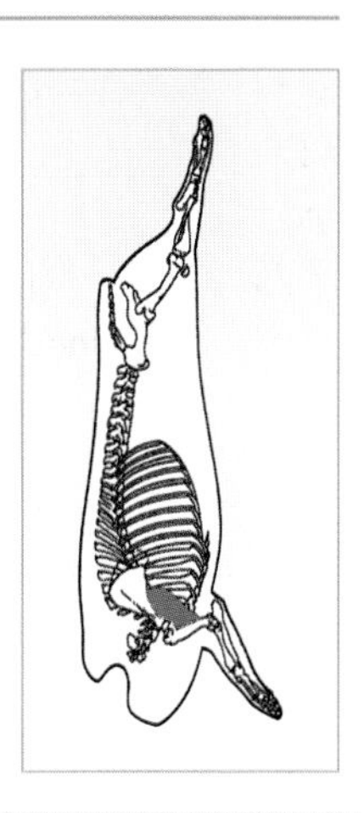

4165-4167 SHOULDER RIBS

Shoulder ribs are derived from a shoulder outside (item 4045) and shall contain three optional levels of trims:

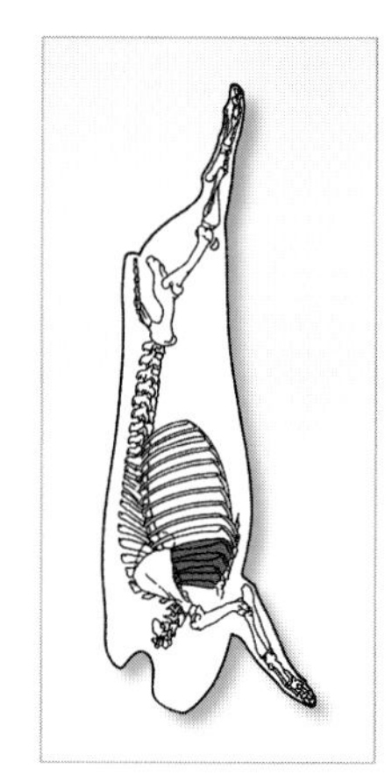

- *No trim* - M. pectoralis profundus retained
- *Marginal trim* - M. pectoralis profundus is trimmed retaining underlying flesh
- *Complete trim* - all lean on the underlying surface of shoulder ribs is trimmed

To be specified:

- Sternum bone retained

ITEM NO.
4165 (4-ribs)
4166 (3-ribs)
4167 (2-ribs)

4350 JOWL

Jowl is removed from the shoulder by a straight cut approximately parallel with the loin side. Jowls shall be reasonably rectangular in shape and at least reasonably squared on the sides and ends. The jowl shall be faced by close removal, of surface glandular and loose tissue, skin and bloody discoloration.

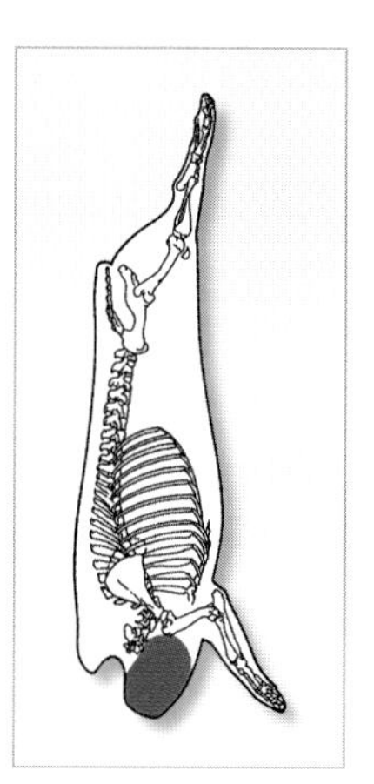

To be specified:

- Skin removed
- Minimum piece size
- Desinewed (exposed heavy (opaque) connective tissue and tendinous ends of shanks removed)

ITEM NO.
4350

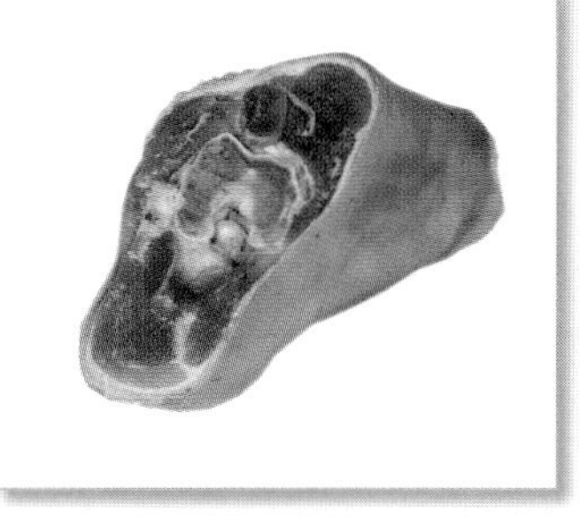

4170 HOCK SHOULDER

Hock shoulder is prepared from a forequarter (item 4021) by the removal of the fore foot at the carpal and radius joints and hock from the shoulder through the radius and humerus bones. The skin shall remain.

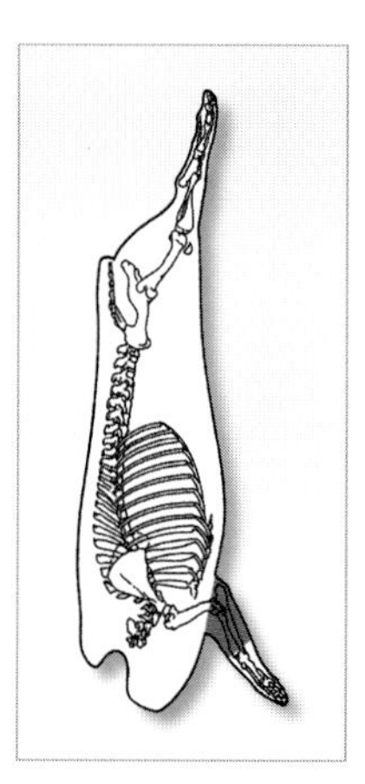

- Skin removed

ITEM NO.
4170

To be specified:
- Skin removed

4175 FORE FEET (TROTTER)

Fore feet (trotter) are prepared from a forequarter (item 4021) at the carpal joint, severing the fore foot (trotter) from the shoulder. The fore feet shall be practically free of hair and hair roots. Skin shall remain.

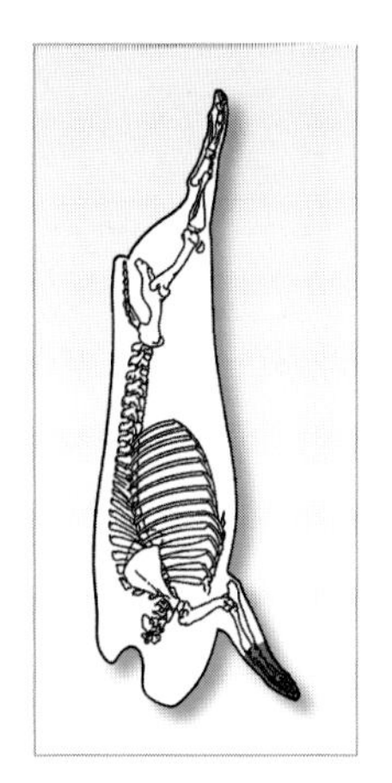

ITEM NO.
4175

To be specified:
- Chemical lean content

4470 TRIMMINGS

Trimmings shall be prepared from any portion of the carcase. Trimmings shall be free of bones, cartilage, skin, seedy mammary tissue and lymph glands (including the prefemoral, popliteal, prescapular and other exposed lymph glands).

7685 BACK FAT

Back fat is prepared from the fatty portion of the back after removal of the loin. Back fat shall be relatively thick and the thickness shall be relatively uniform throughout. All edges must be reasonably squared.

7680 SHOULDER FAT

Shoulder fat is subcutaneous fat prepared from a shoulder.

ANNEX I

CODIFICATION SYSTEM

1. Purpose of the GS1 system

The GS1 system is widely used internationally to enhance communication between buyers and sellers and third-party conformity assessment entities. It is an identification and communication system standardized for use across international borders. It is managed by GS1 Global Office, together with national GS1 member organizations around the world.

The system is designed to overcome the limitations of using company, industry or country-specific coding systems and to make trading more efficient and responsive to trading partners. The use of the GS1 standards improves the efficiency and accuracy of international trade and product distribution by unambiguously identifying trade items, services, parties, and locations. GS1 identification numbers can be represented by data carriers (e.g. bar code symbols) to enable electronic reading whenever required in the trading process.

GS1 standards can be used in Electronic Data Interchange (EDI) and the GS1 Global Data Synchronization Network (GDSN). Trading partners use EDI to electronically exchange messages regarding the purchase and shipping status of product lots. Trading partners use GDSN to synchronize trade-item and party information in their back-end information systems. This synchronization supports consistent global product identification and classification, a critical step towards efficient global electronic commerce.

2. Use of the UNECE code in the GS1 system

GS1 uses application identifiers as prefixes to identify the meaning and format of the data that follow it. It is an open standard, which can be used and understood by all companies in the international supply chain, regardless of the company that originally issued the codes.

The UNECE purchase specification code defined in section 4.1 has been assigned the GS1 application identifier (7002) to be used in conjunction with a Global Trade Item Number (GTIN) and represented in the GS1-128 bar code symbology. This allows the UNECE code information to be included in GS1-128 bar code symbols on shipping containers along with other product information (see examples 1 and 2).

UNECE meat-cut definitions are also being proposed for use by suppliers as an attribute of the GDSN global product classification system. In this way, suppliers can use the UNECE meat-cut code to globally specify the cut of each product GTIN in the GDSN. Once defined by the supplier, all interested buyers will know the exact UNECE cut of each product published in the GDSN (see example 3).

Example 1:

(01)	Global Trade Item Number (GTIN)
(3102)	Net weight, kilograms
(15)	Use-by date
(7002)	UNECE standard code
(10)	Batch number

Example 2:

(01)	Global Trade Item Number (GTIN)
(3102)	Net weight, kilograms
(13)	Slaughter/packing date
(21)	Serial number

Other data, such as the UNECE code, refrigeration, grade and fat depth, can be linked to the GTIN via Electronic Data Interchange (EDI) messages.

3. Application of the system in the supply chain

(1) Customers order, using the UNECE standard and the coding scheme.

(2) On receipt of the order, the suppliers translate the UNECE codes into their own trade item codes (i.e. Global Trade Item Number).

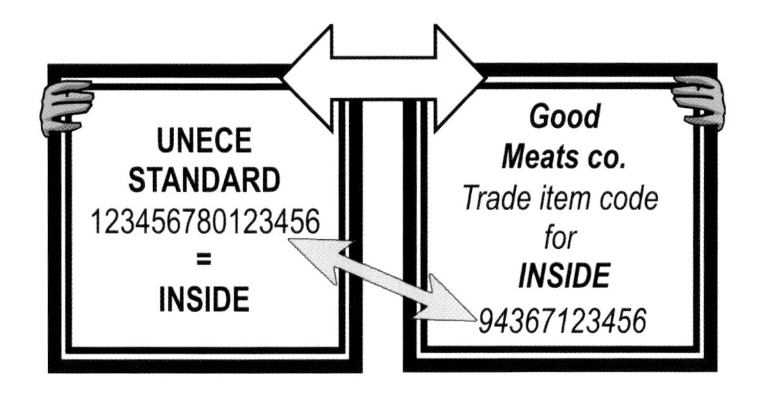

(3) Suppliers deliver the order to the customers. The goods are marked with the GS1-128 bar code symbol.

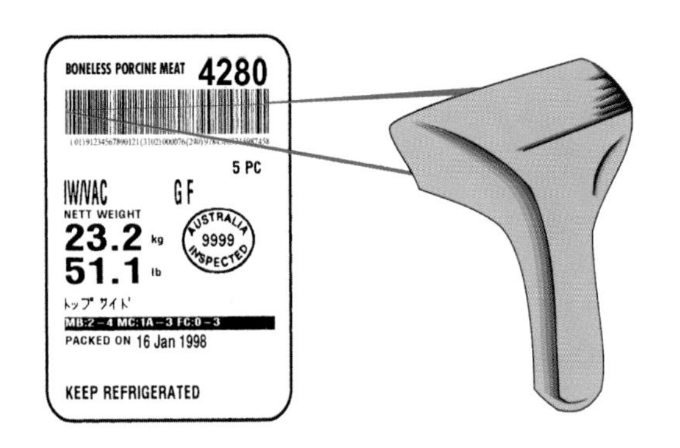

(4) Customers receive the order and the GS1-128 bar code scanned, thus allowing for the automatic update of commercial, logistics and administrative processes.

(5) The physical flow of goods, marked with GS1 Standards, may be linked to the information flow using Electronic Data Interchange (EDI) messages.

Example 3:

4. Use of UNECE meat-cut definitions in the GDSN

(1) Suppliers publish or update information about a product in the GDSN and use the appropriate UNECE meat-cut definition to define the meat cut of the product using the GDSN meat-cut attribute.

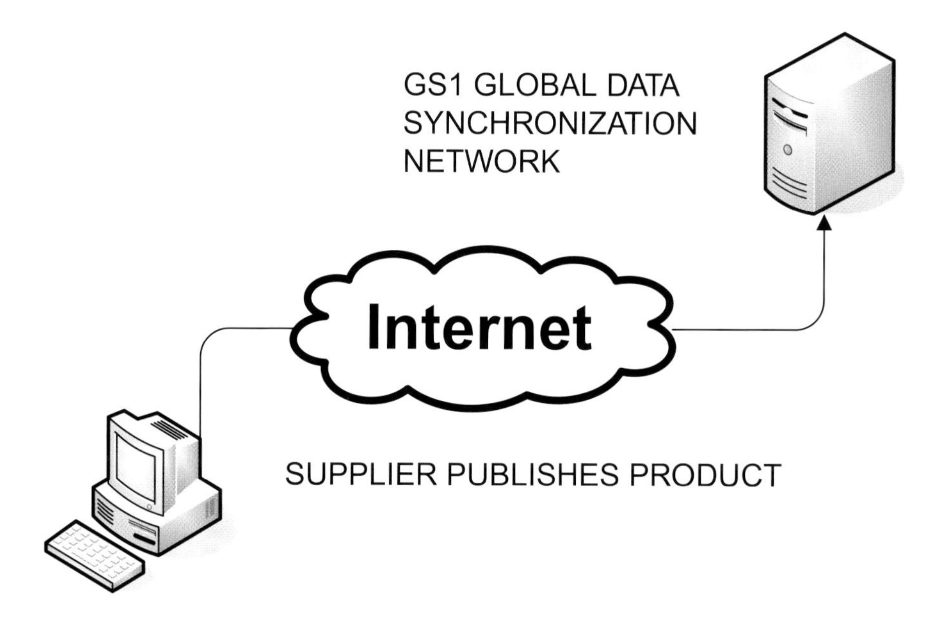

(2) Interested buyers use the UNECE meat-cut and other product information published in the GDSN to synchronize product information in their own information systems.

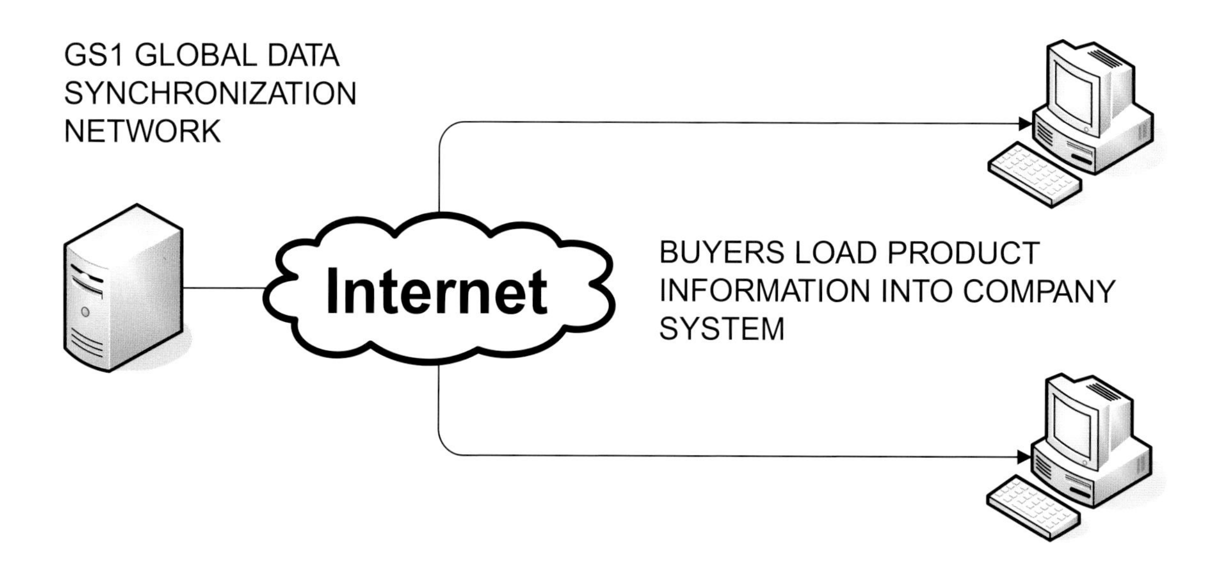

(3) Buyers use UNECE meat cut information in their information systems to identify by GTIN which products they wish to order.

BUYERS IDENTIFY PRODUCTS BY INFORMATION IN COMPANY SYSTEM

GTIN	PRODUCT INFORMATION	
112233123456	FOREQUARTER	4021
112270123457	HIND FEET	4176
998870123001	TENDERLOIN	4280
998870123017	LEG LONG CUT	4013
998870123560	BELLY	4079
776670678444	BELLY	4079
112233123458	LEG LONG CUT	4013
998870123334	FOREQUARTER	4021
776670678427	LEG LONG CUT	4013

(4) Buyers use product GTIN and related information to order product from supplier using EDI or GDSN-compatible data pool service providers.

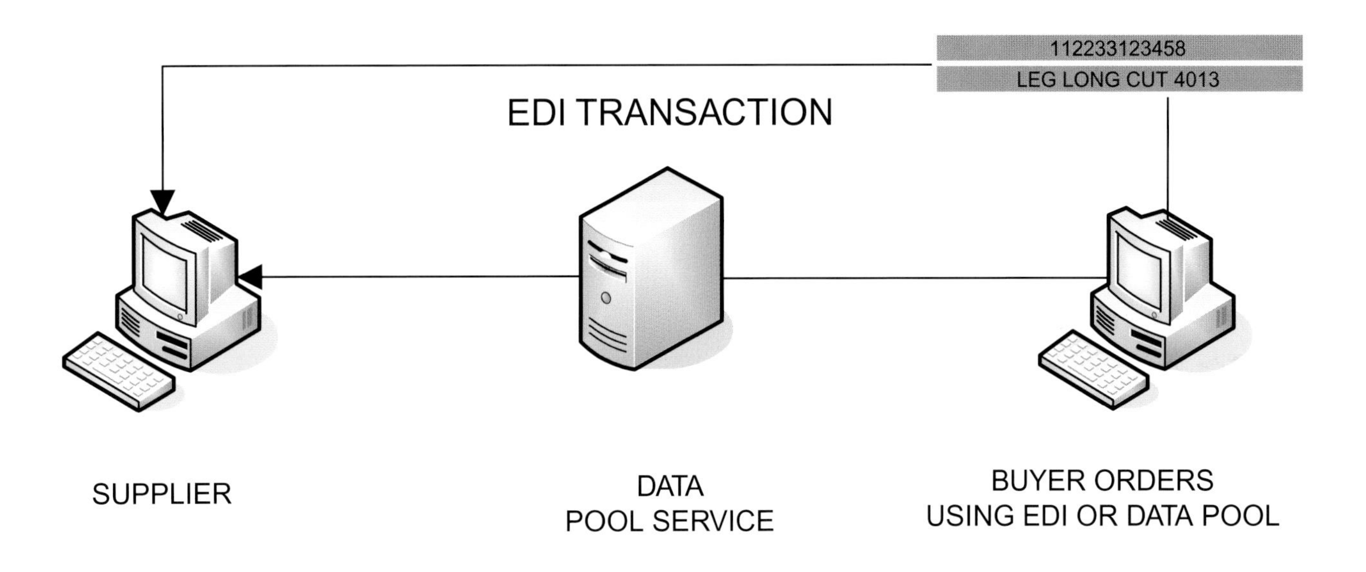

ANNEX II

ADDRESSES

United Nations Economic Commission for Europe (UNECE)
Agricultural Standards Unit
Trade and Timber Division
Palais des Nations
CH - 1211 Geneva 10
SWITZERLAND
Tel: +41 22 917 1366
Fax:+41 22 917 0629
e-mail: agristandards@unece.org
www.unece.org/trade/agr

AUS-MEAT Ltd
Unit 1 / 333 Queensport Road North
Murarrie
Queensland 4172
AUSTRALIA
Tel: +61 7 3361 9200
Fax:+61 7 3361 9222
e-mail: ausmeat@ausmeat.com.au
www.ausmeat.com.au

United States Department of Agriculture (USDA)
Agricultural Marketing Service
Livestock and Seed Program
1400 Independence Ave., S.W.
Washington D.C. 20250 0249
UNITED STATES
Tel: +1 202 720 5705
Fax:+1 202 720 3499
e-mail: craig.morris@usda.gov
www.ams.usda.gov

GS1 International
Blue Tower
Avenue Louise, 326
BE 1050 Brussels
BELGIUM
Tel: +32 2 788 78 00
Fax:+32 2 788 78 99
www.gs1.org/contact/

Designed and printed by Publishing Service, United Nations, Geneva
GE.07-20504–November 2008–1,700

ECE/TRADE/369

United Nations publication
Sales No. E.07.II.E.1

ISBN 978-92-1-116953-9
ISSN 1810-1917